building a recording studio

by Jeff Cooper, M. Arch., S.M., S.B., B.S.A.D.

Consultant in Acoustics
Registered Architect
Jeff Cooper Architects, A.I.A., Los Angeles

Fifth Edition

SYNERGY GROUP, INC.

Los Angeles

Building A Recording Studio
Fifth Edition

published by

SYNERGY GROUP, INC.
23930 Craftsman Road • Calabasas, CA 91302

Printed in the United States of America

Library of Congress Catalog Number: 84-90061
ISBN 0-916899-00-4

to my mother and father,
for their constant love and encouragement

acknowledgements

acknowledgements

There are many people whose efforts have contributed in a special way to this book. I am grateful for these contributions and would like to make the following acknowledgements. I would like to thank Bob Newman, for his guidance and wisdom as my professor at M.I.T.

I would also like to thank the following friends, family members and associates for their numerous contributions ranging from selected editing of the manuscript to overall encouragement and support: Shelley Bernstein, Craig Clarke, June Davis, Jonathan Loring, Garvin Haist, Alex Balster, Susan Feldman, Linda Wheatcroft, Nancy Cooper and Richard Cooper.

contents

acknowledgements vi

preface xi

chapter one Principles of Acoustics 3

what is sound, 3; waves, 3; frequency, 3; velocity, 4; wavelength, 5; amplitude, 6; phase, 7; constructive and destructive interference, 8; musical sounds: harmonics, 8; timbre, 10; shape, 11; harmonic distortion, 12; envelope, 14; octave bands and third-octave bands, 15; resonance, 16; loudness RMS, 17; intensity, 18; the decibel, 19; adding decibels, 20; sound pressure, 21; the decibel: conclusion, 23; the Fletcher-Munson curves, 23; weighted scales: the "A" scale, 25

chapter two How Acoustics Affect Recording 28

introduction, 28; noise, 29; reverberation, 36; absorption, 43; coloration, 50; echo, 53; diffusion, 56; summary, 61

chapter three How to Soundproof a Room 64

introduction, 64; for small budgets: room openings, 65; doors, 67; windows, 69; walls, 72; ceilings, 84; floors, 87; loose objects, 93; appliances, 93; don't add absorption to your room, 94; for larger budgets: introduction, 95; the floating studio: walls, 97; floors, 100; ceiling, 102; doors, 104; windows, 105; ventilation, 107; wiring, 110

chapter four The Studio 116

introduction, 116; location, 117; the floor plan, 119; getting a permit, 119; avoiding parallelism, 120; eliminating echoes, 121; controlling flutter, 122; eliminating standing waves, 123; building the basstrap, 126; the multi-dimensional basstrap, 129;

augmenting the basstrap, 131; eliminating severe colorations, 131; gobos, 132; drum booths, 136; live vs. dead areas, 138; acoustical flexibility, 140; finish materials, 142; lighting, 143; aesthetics, 144

chapter five — The Control Room 148

introduction, 148; the combined studio and control room, 149; the optimum control room, 151: isolation, 152; size, 153; geometry, 154; symmetry, 156; positioning the monitors, 156; the monitor spacing, 158; the monitor path, 160; flush mounting, 160; machine soffits, 162; rear room reflections, 163; control room ceiling, 164; other functions: visitors, 166; interface, 167; troughs, 168; radio frequency interference, 169; tuning the room, 169; evaluating the room, 175

glossary 178

bibliography 200

preface

Some have attributed it to the technological explosion in the science of sound reproduction in the last fifteen years. Others have pointed to the cultural revolution of the 60's (ushered in by the likes of the Beatles and the Rolling Stones), which made music a pivotal force in the lives of so many. Still others have pointed to the new musicians, composers and writers themselves, whose search for artistic freedom has led in directions never considered twenty years ago.

Whatever the reason, the recording studio phenomenon is exploding. In 1960 there were only several hundred professional recording studios in the United States. By 1978 there were over ten times that number. By 1984, the year of this latest edition, semi professional and home studios, which were almost nonexistent in 1960, number an estimated 50,000.

As might be expected, the boom in music and recording has been accompanied by an upheaval in interest in all the peripheral and technological aspects of recording: tape recorders, microphones, mixing consoles and, of course, the science of acoustics.

Paradoxically, although an abundance of excellent information has been available concerning the *equipment* aspects of recording (much of it provided by manufacturers), up to this point very little or nothing has been available about the *acoustical* aspects of recording. This has been truly unfortunate, for the acoustical aspects of the recording environment are equally and sometimes more important than the recording equipment itself, with regards to the ultimate quality of the product, as well as to the enjoyment of the recording process.

As a result of this obvious gap, a tremendous amount of curiosity (an amount which seems to expand expontentially) has been generated about two primary areas of acoustics connected with the playing and recording of music. These are the areas of soundproofing and the design of environments for playing and listening.

In writing this book, I have attempted to address both of these areas. Although the book is, of course, geared towards those building an actual studio, I have tried to organize the information presented so that it will be valuable for an even wider range of acoustical applications. Towards this end, each chapter has been written as a self-contained unit so that it may be independently read by those seeking a specific category of information.

Where possible, solutions have been categorized according to intricacy, which corresponds roughly to cost. I have avoided making specific cost references as these seem to be hopelessly invalid as soon as the publishing date occurs. Furthermore, cost indexes always will vary markedly with the manner in which the work is performed (by contractor, by oneself, etc.), the geographical location of the project, and the seasonal changes in material and labor supply.

In compiling the material, I have avoided an overly theoretical approach which I felt would be inappropriate for those without the background or desire to delve into the more oblique aspects of acoustics. I have tried, however, to provide enough theory and background discussion to make the book interesting for the more knowledgeable reader. Formulas have been referred to only sparingly where I felt they were helpful in augmenting the text.

Of course, there will be those who would have preferred to see all formulas omitted entirely, just as there will be those (more theoretically oriented) who might have enjoyed seeing particular theories explored in even greater mathematical depth. I can only remind the former group that acoustics is, by its very nature, a science, and any approach which abandons all relation to science would be inherently inaccurate. To the latter group, I must emphasize that this book is intended to be practical. The more mathematical and theoretical data they seek is perhaps better suited to another type of book.

Simply stated, the purpose of this book is to provide a library of practical solutions tempered with enough text to make the reading interesting and informative. Accordingly, diagrams and drawings have been used extensively, to facilitate understanding.

The layout has been composed and presented in a manner which is designed to make reading an enjoyable experience. A glossary of technical terms used in the book has been provided at the back for quick working definitions. Lastly, an appendix dealing with the basic principles of acoustics has been included for those who find themselves in unfamiliar territory.

Among the new inclusions in this fourth edition are several updates covering new developments in control room design. Feedback I have received from the readers of earlier editions indicates a desire for even more detailed information regarding the design of room treatments and basstrapping. Therefore, additional formulae have been added to the text to enable more accurate calculation of basstrap design for specific situations.

In closing, I sincerely hope that, through the provision of information and perhaps technical understanding, I have in some small way encouraged the recording and dissemination of music, a force whose vast energies I feel are only beginning to be tapped.

Jeff Cooper
Los Angeles, 1984

chapter one

principles of acoustics

introduction

Building A Recording Studio is intended to be a practical reference manual and workbook. It is divided into self-contained chapters to make specific types of information readily accessible.

The first chapter, the Principles of Acoustics, provides a grounding in the basic principles for those encountering the subject for the first time or for those seeking a refresher course.

The second chapter, How Acoustics Affect Recording, relates these principles to the recording process and explores such acoustical concepts as noise, absorption, and reverberation.

Chapter Three, How To Soundproof A Room, is a guide to the principles of soundproofing. It contains many tried and tested solutions to the universal problem of soundproofing. Although geared toward studio applications, the information provided in this chapter can be applied to almost any other building type, including a home, apartment, or work environment, since the basic construction principles remain the same.

Chapter Four, The Studio, outlines the function of a recording studio, whether for music, voice, drama or other purposes and presents suggestions for room layout, geometry, and interior treatments. Construction diagrams for such items as drum booths, baffles, and basstrapping are provided.

Chapter Five, The Control Room, describes the principles of room acoustics as they relate to listening in a controlled environment. Tested and proven construction techniques are outlined. This chapter also contains updates on recent innovations in the area of control room design.

what is sound?

Sound is the sensation caused by the vibration of the ear drum. When an object or *source* is caused to vibrate, a series of *pressure impulses* is created in the surrounding air. As these impulses strike the ear drum they are translated into minute electrical signals which are, in turn, sent to the brain for deciphering.

waves

When a source such as a guitar string is set into vibratory motion by plucking, the surrounding air reacts in an *elastic* manner. It is alternately *compressed* (squeezed) and *rarefied* (expanded) by the rapid back-and-forth movement of the vibrating object. One complete cycle of compression and rarefaction is called a *wave*.

As successive waves are formed, they travel outward in a spherical pattern, in much the same way that water waves travel outward from a plunging stone.

frequency

The speed with which the waves are regularly produced is called the *frequency* (f) of vibration and is measured in *cycles per second* (c.p.s.) or *hertz* (Hz). The length of time taken for one complete cycle is called the *period* (t) of vibration. Humans can hear frequencies in the approximate range of 20 to 20,000 Hz. This range is often referred to as the *audible spectrum*.

The frequency of a sound is related to the *pitch* heard by the ear. In general, the higher the frequency, the higher

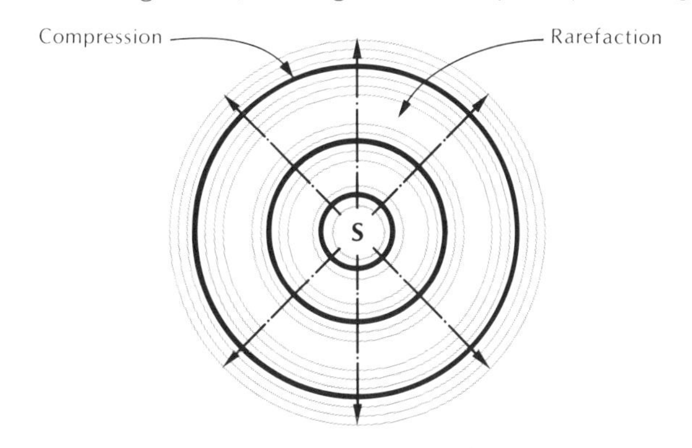

Figure 1-1 Sound waves generating from a source

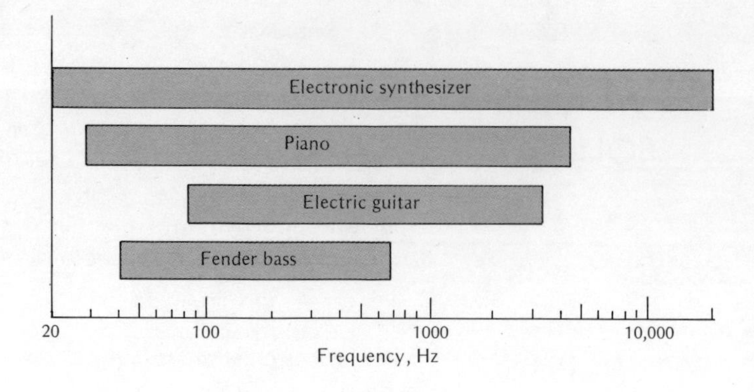

Figure 1-2 The frequency ranges of several instruments

the pitch. The low note of a Fender bass corresponds to a frequency of 41 Hz, while the open high string of an electric guitar corresponds to 330 Hz. The highest note of a piano is 4,186 Hz. Some electronic music synthesizers can produce tones all the way up to 20,000 Hz.

velocity

In air, at 70°F (21.1°C), sound-pressure waves travel outward from the source at 1,130 feet per second. This is commonly referred to as the *velocity* of sound. This velocity varies slightly with air temperature and humidi-
In *solid* materials such as wood and steel, sound travels much more quickly (11,700 and 18,000 feet per second. respectively). These materials are more *elastic* than air (i.e., their molecules tend to bounce back more quickly

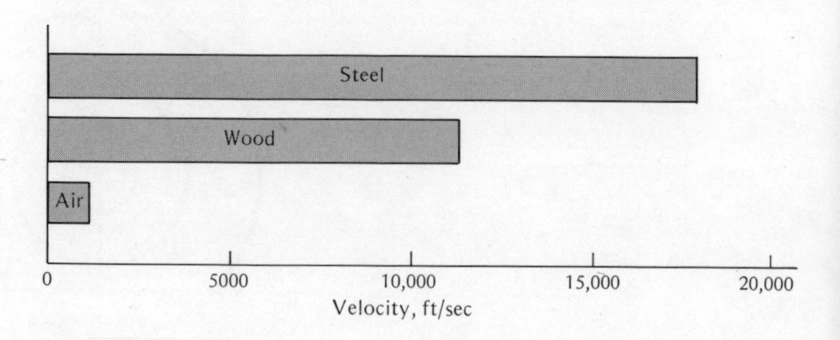

Figure 1-3 Velocity of sound in different materials

when deflected by a sound source). The result is that a vibrational impulse is easily sent scurrying through wood and steel, accounting for the remarkable ability of most buildings to conduct sound directly through their structural framework.

wavelength

The absolute distance that a wave travels to complete one full cycle of compression and rarefaction is called the *wavelength*, and is represented by the symbol λ

Velocity, frequency and wavelength are related by the following formula:

$$v = f(\lambda)$$

Since velocity of sound in air is constant, the frequency and wavelength change inversely with one another. The higher the frequency of a sound, the shorter its wavelength will be.

Thus, for a tone of 40 Hz., $\lambda = \frac{v}{f} = \frac{1130}{40} = 28.25$

1,000 Hz., $\lambda = \frac{v}{f} = \frac{1130}{1{,}000} = 1.13$

20,000 Hz., $\lambda = \frac{v}{f} = \frac{1130}{20{,}000} = .0565$

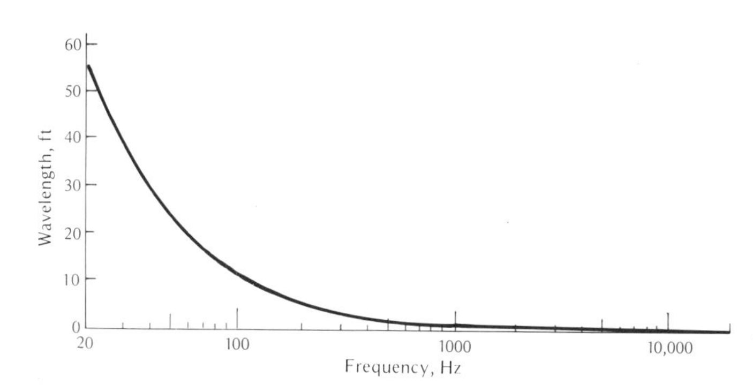

Figure 1-4 Several frequencies and their corresponding wavelengths

amplitude

In the simplest sound wave, that of a pure tone, air pressure in the wave varies alternately above and below normal atmospheric pressure in accordance with the following diagram. (Normal atmospheric pressure is represented by the zero line.)

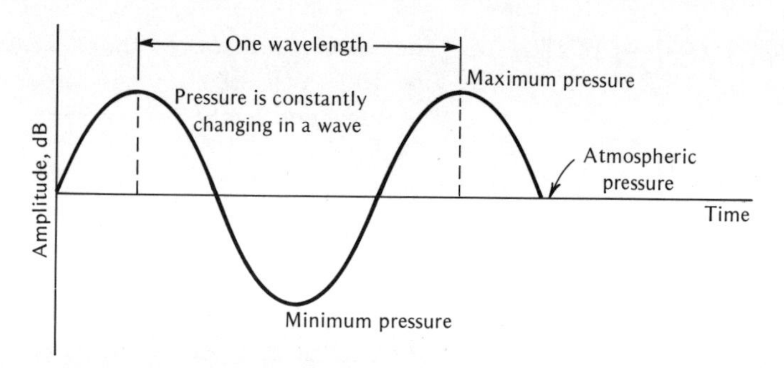

Figure 1-5 A simple wave

The amount that the pressure varies above and below normal is represented by the distance of the curve above or below the zero line. This distance, at any given time, is referred to as the *amplitude* of the wave and corresponds roughly to the loudness of a sound. The amplitude of a pure tone, such as that pictured, follows the trigonometric sine function. Any wave that repeats and maintains its symmetry about the zero line, such as this, is called a *simple* waveform. Waves that do not necessarily repeat or show symmetry above and below the zero line are called *complex waves*. Banging on a trash bin, for

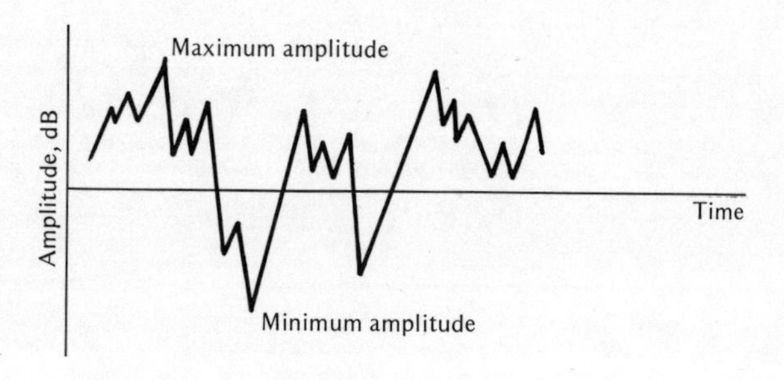

Figure 1-6 A complex wave

example, would produce a complex waveform. A spoken word also would produce a complex waveform.

phase

Since waves are repetitive phenomena, they can be divided into regularly occurring intervals. These intervals are normally measured in *degrees*.

Each full wavelength corresponds to 360 degrees of travel. Therefore, one-half a wavelength corresponds to 180 degrees. In a simple wave, the maximum amplitude occurs at one-quarter wavelength, or 90 degrees.

Two waves that commence their excursions at the identical moment are said to be *in phase*. (This includes waves that are a full 360 degrees apart.) Waves that commence their excursions at any other time are said to be *out of phase*.

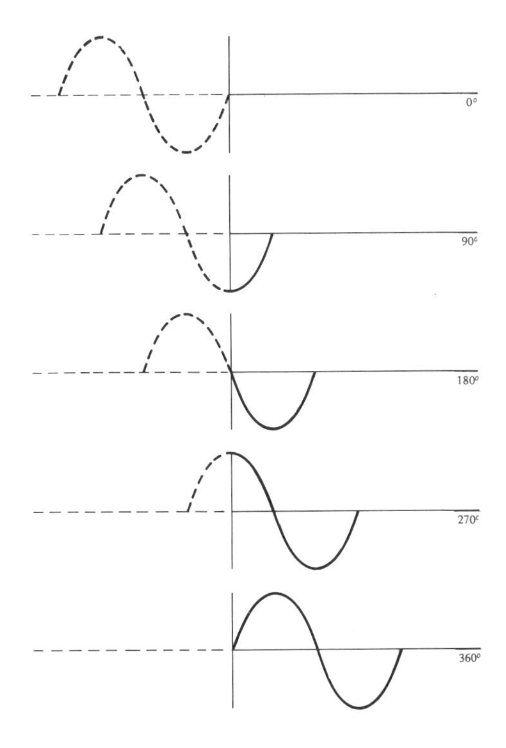

Figure 1-7 Phase relationships of five identical waves

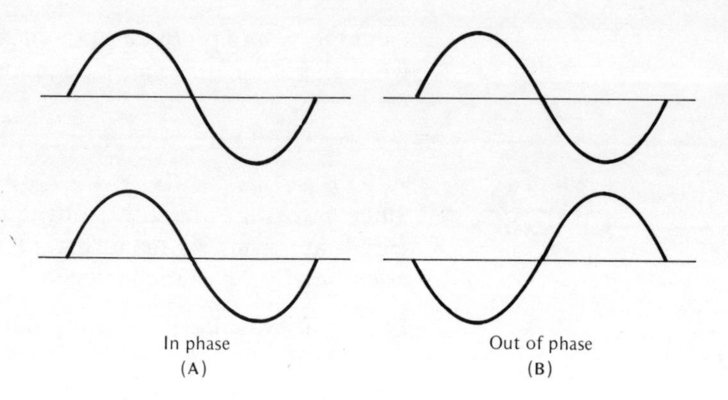

Figure 1-8 An illustration of phase relationship

constructive and destructive interference

Sound waves, being forms of energy, can be readily added to one another. The result, however, is not always greater than both of the parts.

The *phase relationship* of the constituent waves plays a great deal of importance in determining how two or more waves will add. Consider the case of two identical waves, starting at the same instant from the same point in space. It is easy to visualize how these waves would increase each other's absolute amplitude at all points. On the other hand, if the second wave were to begin its positive excursion at the same moment the first wave were beginning its negative excursion, it is not difficult to see that the second wave might actually cancel out the first wave entirely.

The intersection of two or more wavefronts that results in a greater positive amplitude than either of the constituent waves is called *constructive interference*. The intersection of two wavefronts that results in a greater negative amplitude than either of the constituent waves is called *destructive interference*.

musical sounds: harmonics

There is a distinct relationship between a frequency and its integral multiples (i.e., between 200 Hz, 400 Hz, 600 Hz, 800 Hz, etc.). These multiples are called *harmonics*. The *fundamental* tone (200 Hz) is referred to as the first harmonic. Successive multiples are referred to as the second harmonic (400 Hz), third harmonic (600 Hz), fourth harmonic (800 Hz) and so on.

Most people hear a recognizable musical relationship between a frequency and its multiples. The first harmonic or double is called a *musical octave*. The second harmonic is a musical fifth and is related to the first and second harmonics in a unique and recognizable way. This pattern of harmonic relationships is true for all musical notes, regardless of fundamental frequency.

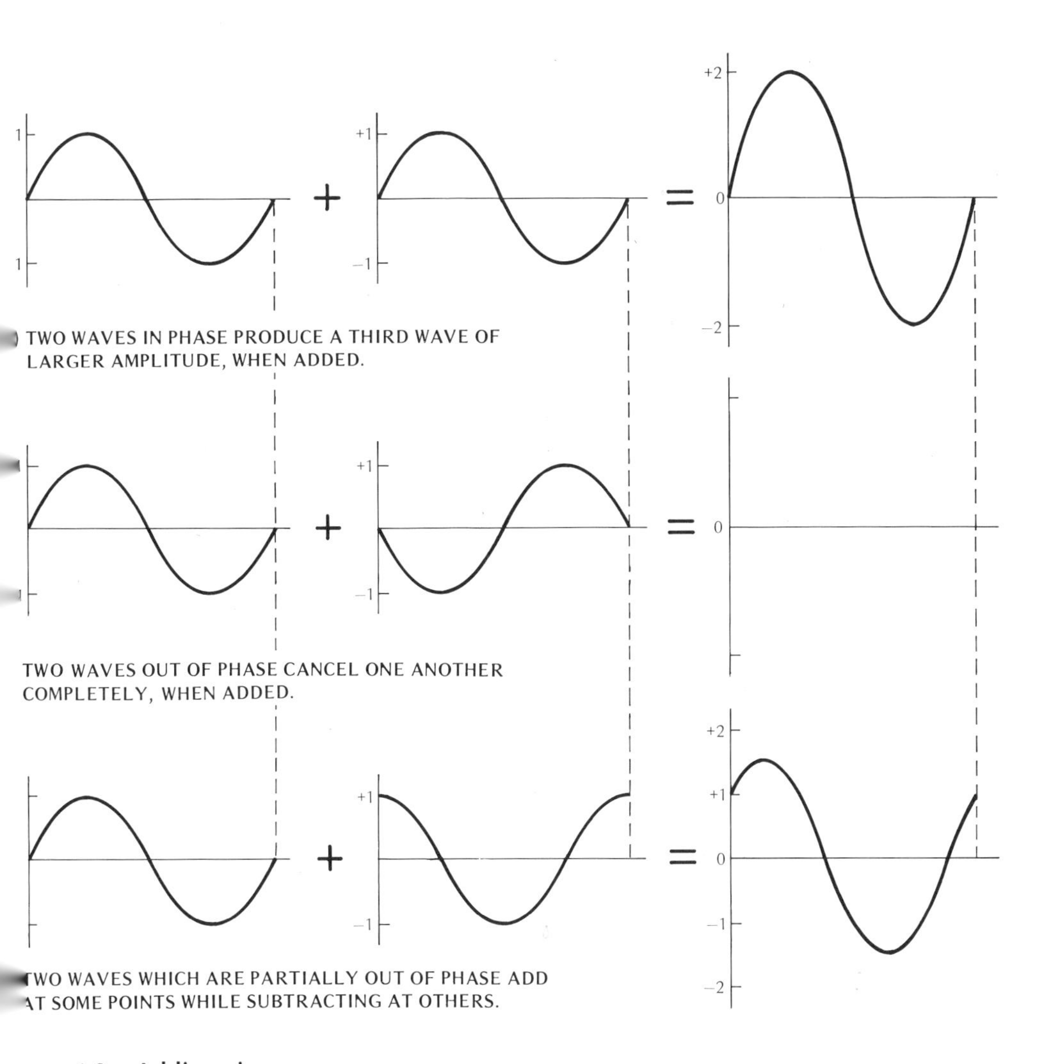

ure 1-9 Adding sine waves

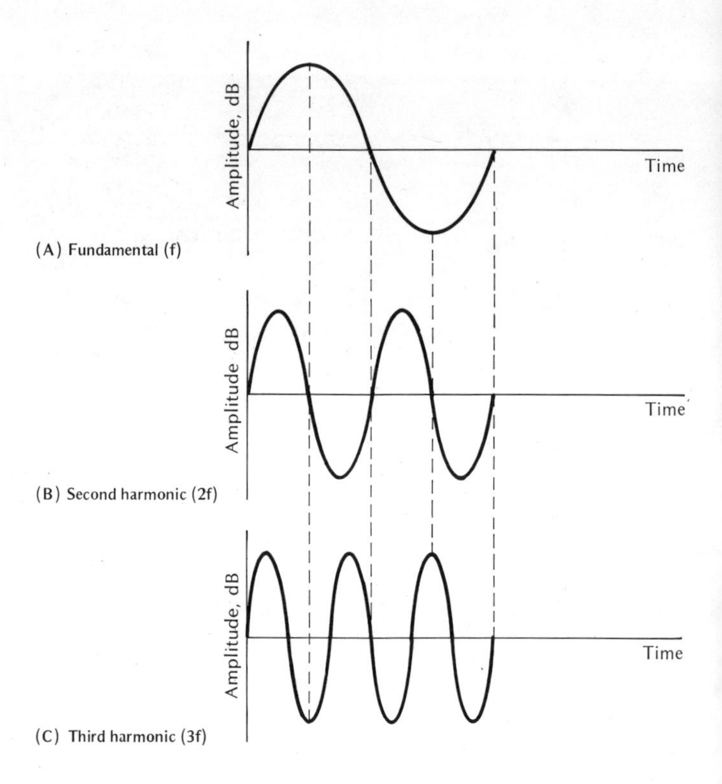

Figure 1-10 The harmonics of a simple wave

timbre

Any note produced by a musical instrument contains varying components of the fundamental frequency and its harmonics. It is actually the different combinations and prevalences of these harmonics that are responsible for differences in *timbre* or tone between various musical instruments. Thus, a saxophone note at 440 Hz (A) sounds characteristically different from a trumpet tone of the same fundamental frequency because the harmonic structures produced by these instruments are radically different. (The difference in sound also results from the difference in their wave shapes and envelopes, which will be discussed in the next sections.)

Each harmonic or set of harmonics adds a characteristic acoustical color to the sound. For instance, the second harmonic adds fullness. The third, or musical twelfth, seems to *cover* or soften the sound. Harmonics above the seventh harmonic are called *edge harmonics* and give *bite* and definition to the tone.

In general, *even*-numbered harmonics (second, fourth, sixth, etc.) *open up* or *fill out* the sound. *Odd*-numbered harmonics *close off* or *stop down* the sound.

shape

The harmonics present in a sound wave give its waveform a characteristic *shape*. The waveform of a simple fundamental tone has the shape of a sine wave, as depicted earlier. The addition of a second harmonic creates a new, more complex shape. Similarly, the addition of a third harmonic creates an even more intricate waveform. In general, the shape of any waveform is determined by the complexity of its harmonic structure.

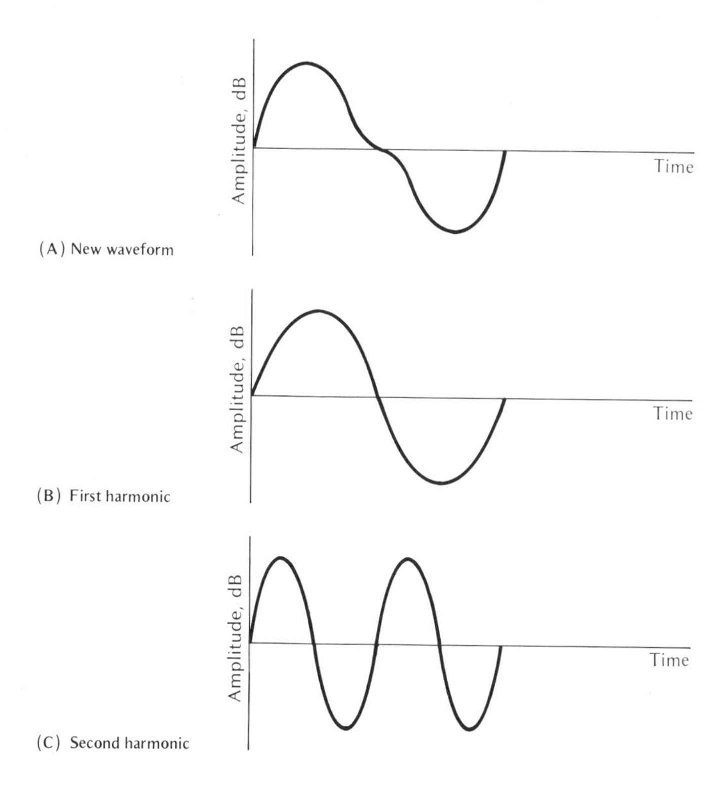

Figure 1-11 Adding harmonics to produce a new waveform

It has been mathematically shown that *all* waveforms, regardless of their complexity, can be produced by the addition of a fixed combination of sine waves of correct amplitude, frequency and phase.

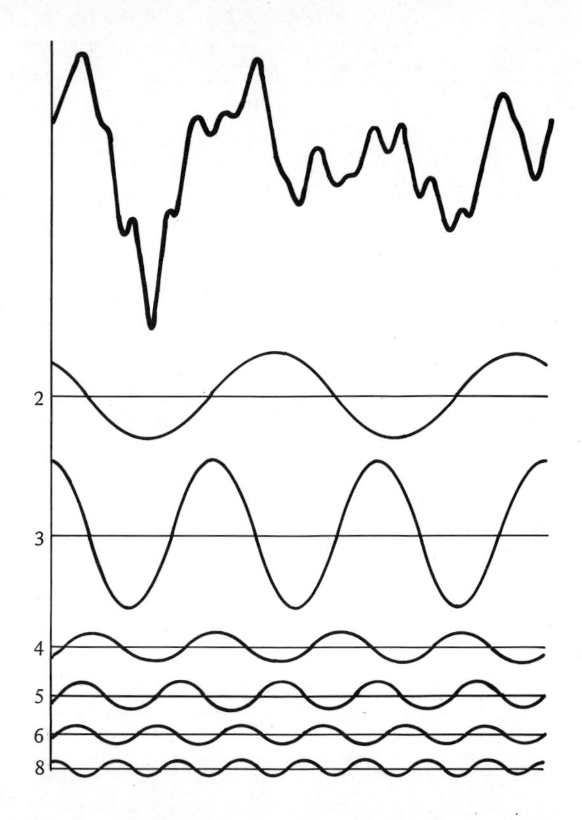

Figure 1-12 The shape of a waveform is determined by its harmonic content

harmonic distortion

To complicate matters, the ear adds its own harmonics to sounds at loud listening levels. Thus a 500-Hz tone, played loudly enough, will produce (among others) a 1,000-Hz and 1,500-Hz harmonic in the ear. In addition, edge harmonics, which would normally go unnoticed, become emphasized. This production of harmonics not present in the original signal is called *harmonic distortion.* This effect explains why some musical sounds take on a harsh characteristic when played at loud listening levels.

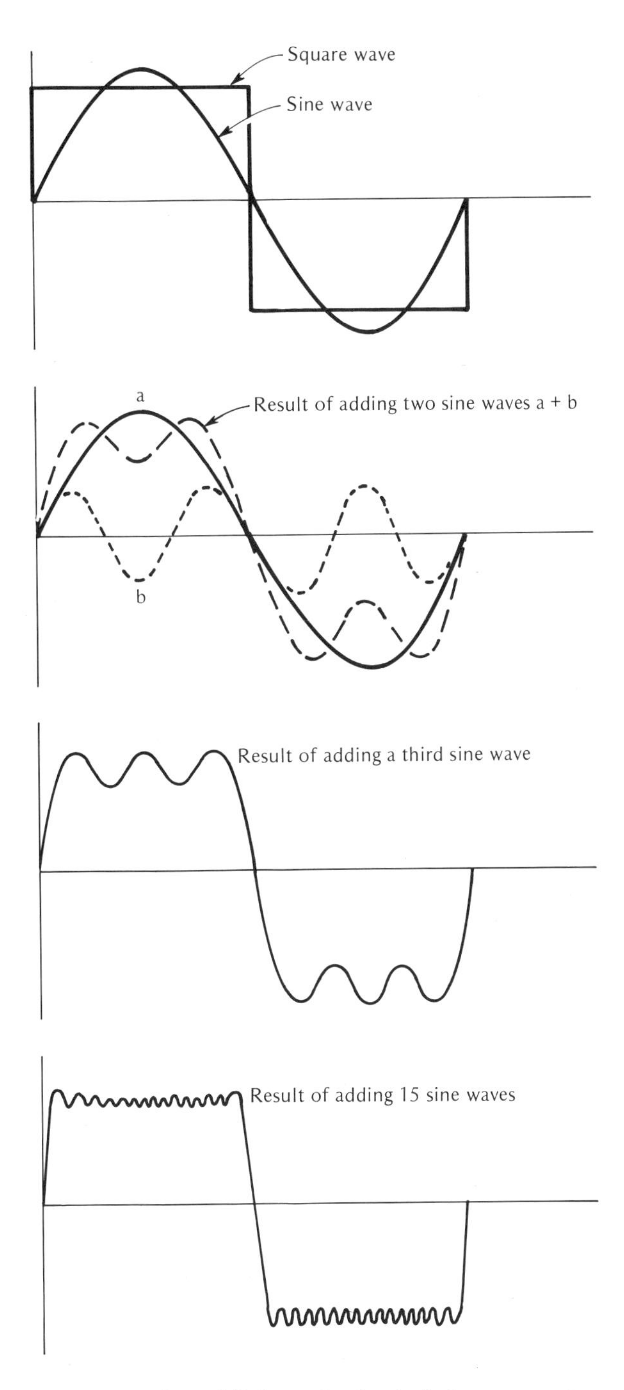

Figure 1-13 The addition of selected sine waves to produce a square wave

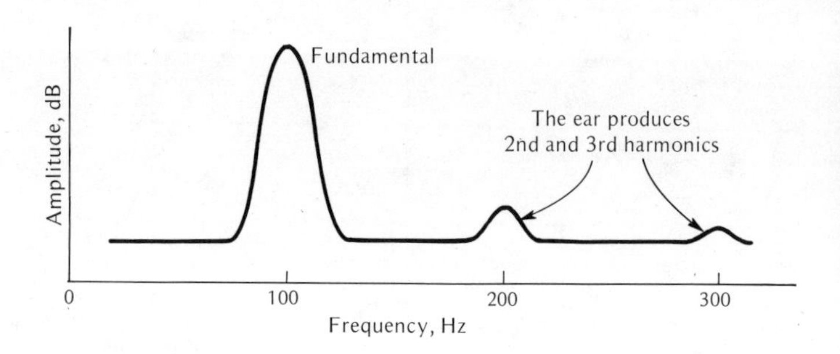

Figure 1-14 The production of harmonic distortion

envelope

Most sounds have a definite start, a limited duration and a period of decay. Thus, the waveform of a sound is often divided into three sections to correspond to these three periods. These sections are called the *attack* (the period in which the sound starts up), the *internal dynamics* (the period in which the sound endures) and the *decay* (the period in which the sound eventually dies away).

The *envelope* of a waveform graphically illustrates the general behavior of a sound during these three periods. An organ tone might have a fairly fast attack, a long sustain and a gently sloping decay. The result would be a simple three-section envelope.

Most sounds are more complicated, possessing several periods of decay and sustain. Percussive waveforms (such as those from a snare drum), for example, have steeply rising attacks and steeply falling *initial decay,* after which the internal dynamics (i.e., the resonance inside the drum) persist and decay at their own rate.

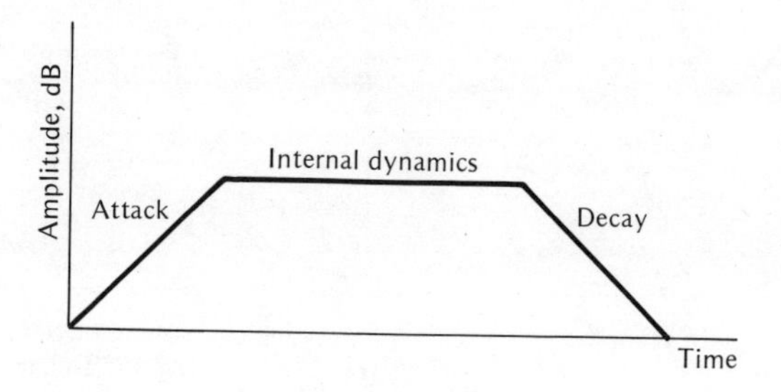

Figure 1-15 The basic envelope of a waveform

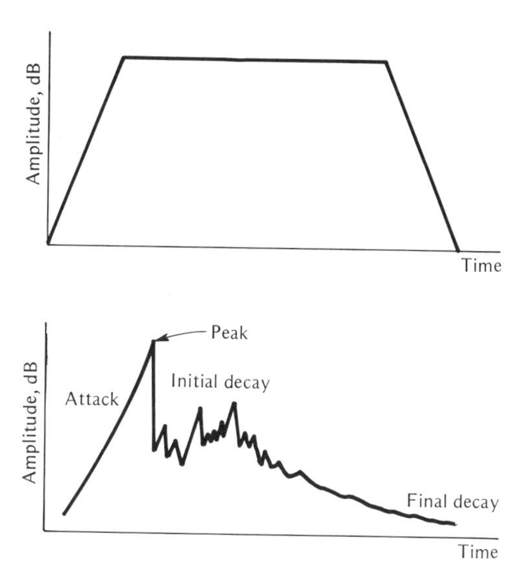

Figure 1-17 Envelope of a cymbal crash

octave bands and third-octave bands

Up until now, we have been chiefly concerned with single notes and their overlying harmonics. Most musical sounds are actually much more complicated than this, possessing dozens of fundamental frequencies and hundreds of harmonics (piano chords, guitar chords, group sounds, etc.).

To enable us to visually represent this potpourri of sounds, we usually resort to a *frequency-spectrum graph.*

The audible frequencies are shown, from left to right, on the horizontal axis. The prevalence of each particular frequency is measured on the vertical axis. Since the thousands of actual frequencies cannot be individually listed in a single diagram, a compressed scale of frequencies is often used (i.e., a *logarithmic* scale). This compressed scale is usually separated into narrow divisions or *bands.*
For *musical* analysis, the spectrum is often divided into *octave bands.* For acoustical analysis, more accurate scrutiny is required. Thus, the spectrum is usually di-

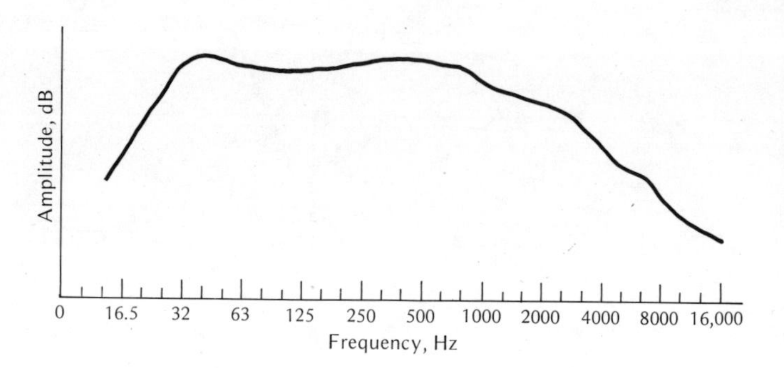

Figure 1-18 Frequency analysis by third octave bands

vided into *thirds* of octaves or *third-octave bands*. The octaves centered at the following frequencies have become the conventional divisions: *16.5 Hz, 31.5 Hz, 63 Hz, 125 Hz, 250 Hz, 500 Hz, 1,000 Hz, 2,000 Hz, 4,000 Hz, 8,000 Hz, 16,000 Hz.*

resonance

All objects, regardless of size, have a natural or resonant frequency, determined partially by their mass and stiffness. A crystal glass will hum and finally shatter if induced to resonate at its natural frequency. In general, if an object is approached by a sound in the vicinity of its natural frequency, it will vibrate. This phenomenon is called *sympathetic resonance.* Sympathetic resonance can be a powerful acoustical phenomenon. In 1940, the

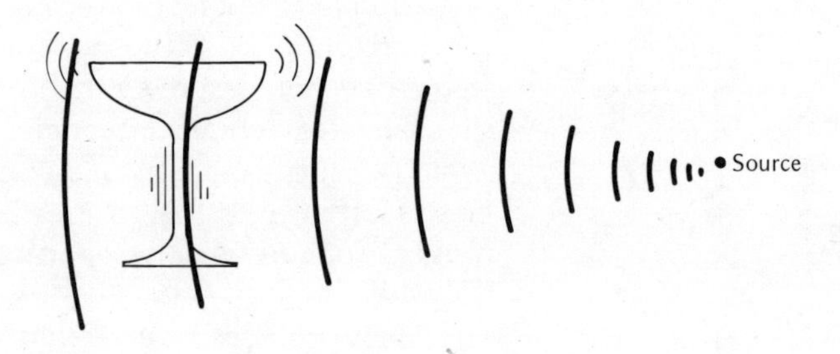

Figure 1-19 Sympathetic resonance of a champagne glass

Tacoma Narrows Bridge, in Tacoma, Washington, was induced to resonate at its natural frequency by a combination of tremendous wind and water-current energy. The bridge began to twist under the immense power of its own vibration and ultimately destroyed itself.)

The principle of sympathetic resonance plays a tremendous role in musical instrument design, as well as studio design.

ɔudness RMS

Although the amplitude of a simple waveform is endlessly fluctuating above and below zero, the sound which the wave produces in the ear is relatively constant. It follows that perceived loudness is not determined exactly by the instantaneous amplitude, but in some other way. Experimentation has shown that the ear gauges loudness of a waveform by *energy content* rather than by instantaneous amplitude. Energy content is related to the *root mean square* (commonly called *rms*) value of the wave.

The rms value derives its name because it is arithmetically computed by taking the square root of the mean values of a waveform after all positive and negative values have themselves been squared. The rms value of a fluctuating wave may be thought of as the value of a non-fluctuating or continuous signal producing an equivalent power.

In an electrical analogy, the rms value of an alternating current equals the value of direct current necessary to produce a similarly powerful signal.

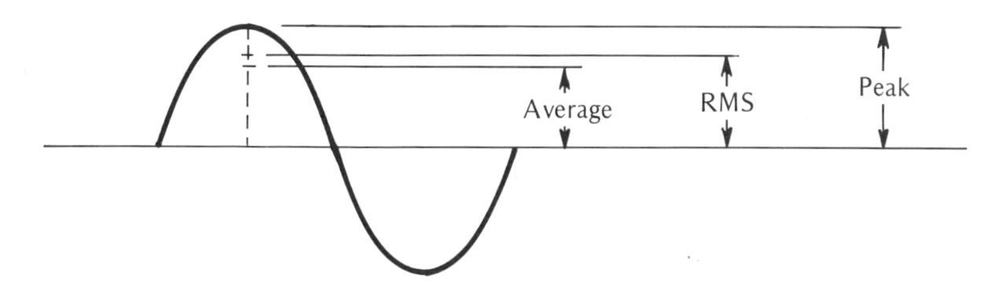

Figure 1-20 The relationship between peak and RMS value of a wave

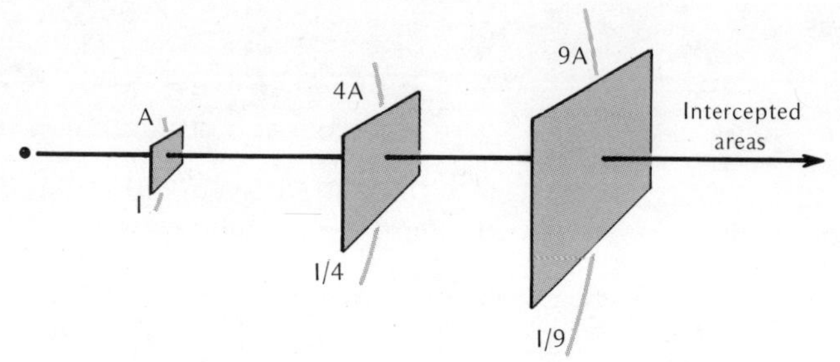

Figure 1-21 Intensity decreases rapidly with distance

intensity

At any given distance from a sound source, the sound wave has a characteristic *intensity*. This intensity decreases rapidly as the wave travels outward. Specifically, intensity diminishes according to the square of the distance travelled. (At twice a given distance from the source, intensity would be *four* times less. At three times the distance, the intensity would be *nine* times less.)

The intensity of a sound field is defined as the flow of acoustic power (measured in watts) through a given area (measured in square meters). Thus, sound intensity is measured in *watts per square meter*.

SOURCE	INTENSITY (watts/m^2)
Rock band on stage	10
Acoustic guitarist (at 3′-0″)	0.000 01
Conversation	0.000 001
Whisper	0.000 000 000 1

Figure 1-22 Various intensities of common sounds

For a sound to be heard, it must have a minimum intensity of .000,000,000,001 watts per square meter (10^{-12}). This is called the *threshold of hearing.*

The human ear is sensitive to a tremendous range of intensities (one-million billion, approximately). These vary from the faint sound of breathing at .000,000,000,001 watts per square meter to the roar of an Apollo moon rocket at 1,000 watts per square meter.

the decibel

If we denoted everyday sounds by the range of intensities they produced, we would need a scale with well over a million-billion values. Such a system would be next to impossible to use. Therefore, a system has been invented to compress the vast range of sound intensities into a more manageable scale. It is called the *decibel system.*

In the decibel system, all intensities are considered multiples of the threshold of hearing. *Each multiple (of ten) is called a Bel,* named after Alexander Graham Bell.

Consider an example: If the intensity of a given sound were three multiples of ten ($10 \times 10 \times 10$), i.e., one thousand times greater than the threshold of hearing, its intensity level would be denoted as 3 Bels. If the intensity were six multiples of ten ($10 \times 10 \times 10 \times 10 \times 10 \times 10$) or one-million times greater than the threshold of hearing, its intensity level would be denoted by 6 Bels.

Since the Bel is a large unit, it is normally divided into ten sub-units. Appropriately each tenth of a Bel is called a *decibel* (abbreviated dB). The first example cited above (3 Bels) would therefore have an intensity level of 30 dB. The second example (6 Bels) would have an intensity level of 60 dB.

The formula for converting intensities to decibels is as follows:

$$dB = 10 \log_{10} \frac{I}{10^{-12}}$$

where

dB = the number of decibels
I = the intensity of the sound being converted, watts/sq. meter
10^{-12} = the threshold of hearing, watts/sq. meter

* *(The logarithm to the base 10 is the mathematical operation that expresses the intensity ratio as a* power, *or multiple of ten, thus converting it to Bels. The factor of ten is introduced into the equation to convert Bels into decibels. By utilizing published tables of logarithms, it is possible to calculate decibel conversions for all intensity ratios, including those that are not exact multiples of ten).*

adding decibels

Since different decibel levels represent different ratios, decibels cannot be added directly to one another. For instance, one sound source, such as a guitar player, might have an intensity level of 70 dB at a distance of several feet. Two guitar players would *not* measure 140 dB. Their total level in dB would have to be calculated by adding the original *intensities.*

i.e., since $70\text{ dB} = 10 \log_{10} \frac{(10^{-15}\text{ watts/sq. meter})}{(10^{-12}\text{ watts/ sq. meter})}$

For two guitarists:

$$\text{dB} = 10 \log_{10} \frac{2 \times 10^{-5}}{(10^{-12}\text{ watts/sq. meter})} = 73\text{ dB}$$

Thus, while one guitar player produces an intensity level of 70 dB, two produce an intensity level of only 73 dB. Four would produce a level of 76 dB, and so on.

In general, *doubling the intensity of sound increases its level by approximately 3 dB.* A sensitive human ear can detect changes of approximately 1 dB. Below that differential, changes are largely imperceptible.

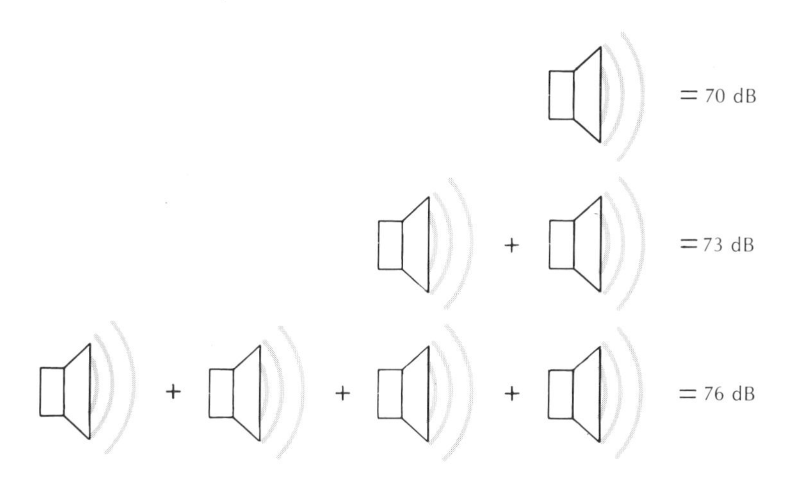

Figure 1-23 Adding decibels

sound pressure

Up to this point, we have discussed the decibel as a tool for comparing the *intensity levels* of various sounds. Converting intensity levels to decibels poses no great problem as long as the intensity of the sound being measured is readily available. This is seldom the case, however.

Intensities of everyday sounds are extremely difficult to obtain. They cannot be looked up in a chart or measured with a simple instrument. This is because intensity represents a mathematical ideal more closely than it does a physical quantity.

However, *pressure* is fortunately related to intensity and *can* be measured using a relatively simple device: the microphone. Therefore, most acoustical measurements given in dB are actually *sound-pressure levels,* rather than intensity levels. The measurements are taken with a microphone and an instrument that electrically calibrates the output of the microphone in dB. Such an instrument is called a *sound-level meter.*

In order that decibels of pressure be numerically equal to decibels of intensity, the pressure value measured must be referenced to the threshold of hearing, and the

INTENSITY (watts/m^2)	SOUND PRESSURE (microbars)	SOUND PRESSURE LEVEL (dB)
100 000 000	2 000 000	200
10 000 000		190
1 000 000	200 000	180
100 000		170
10 000	20 000	160
1 000		150
100	2 000	140
10		130
1	200	120
0.1		110
0.01	20	100
0.001		90
0.000 1	2	80
0.000 01		70
0.000 001	0.2	60
0.000 000 1		50
0.000 000 01	0.02	40
0.000 000 001		30
0.000 000 000 1	0.002	20
0.000 000 000 01		10
0.000 000 000 001	0.000 2	0

Figure 1-24 Intensities, sound pressures and decibels

result must be squared. Therefore, the formula for decibels using pressure is:

$$dB_{(spl)} = 10 \log_{10} \frac{P^2}{.0002}$$

where:

$dB_{(spl)}$ = the number of decibels (spl)
P = the pressure of the sound being converted
.0002 = the threshold of hearing, measured in microbars of pressure

When the term decibel *is used in this book, it will be assumed to refer to sound-pressure level unless stated otherwise.*

the decibel: conclusion

The decibel is probably the most misunderstood and misused term in acoustics. This stems from the fact that it does not really measure a physical quantity. The decibel merely represents a ratio between two quantities.

As such, the decibel can be, and is, used for comparing many physical quantities whose values span a vast range—such as electrical power, sound pressure or even voltage.

Thus, for clarification, the term *dB* should be followed by an abbreviated reference, describing the quantities being referred to. Sound-pressure levels, for example, are denoted as dB (spl).

the Fletcher-Munson curves

Human ears are most sensitive to sounds between 1,000 Hz and 4,000 Hz. Above and below those approximate frequencies, a tone must be several dB greater in order to be perceived as equally loud as a tone in the 1,000- to 4,000-Hz range.

The researchers Fletcher and Munson plotted these varying sensitivities of the human ear across the audible frequency spectrum. Their results have become the stan-

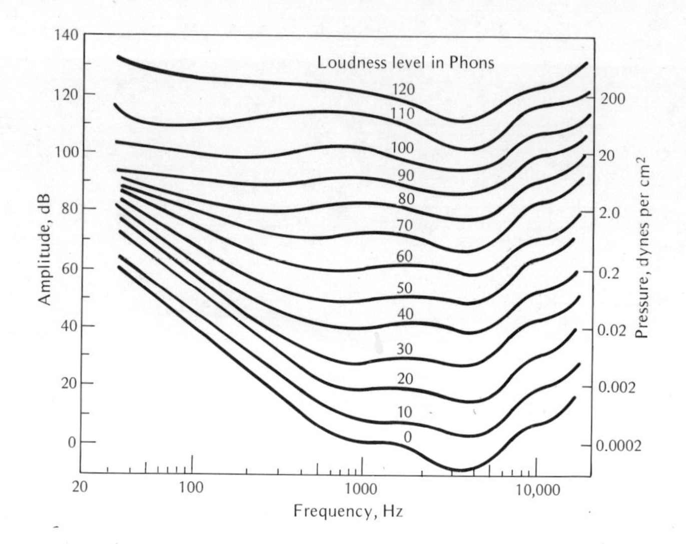

Figure 1-25 The Fletcher-Munson equal loudness contours

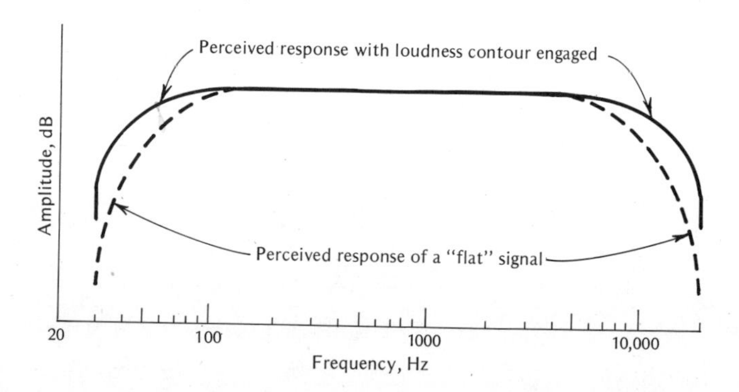

Figure 1-26 The "loudness contour" control of a stereo receiver boosts the low and high ends of the spectrum to compensate for the ear's uneven response at low listening levels

dard reference curves. These curves depict the sound-pressure levels at various frequencies required to produce sensations of equal loudness in the human ear. Referring to the graph, for example, a 40-Hz tone must measure approximately 77 dB (spl) to be gauged as loud as a 400-Hz tone that measures only 50 dB (spl). As can

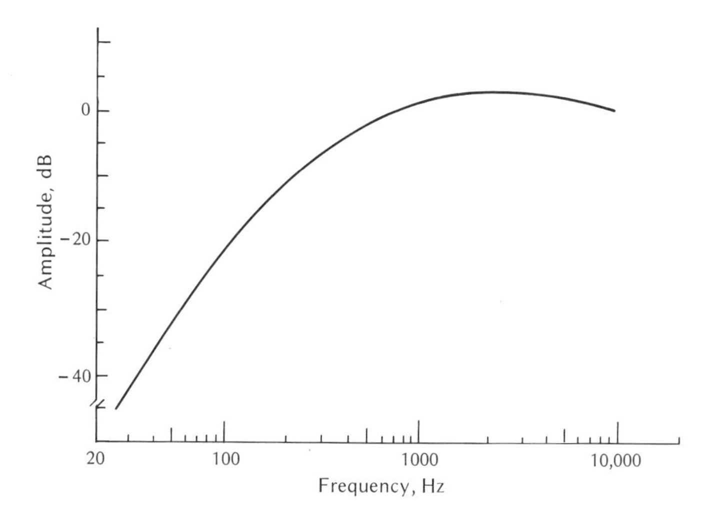

Figure 1-27 The "A" scale simplifies measurement-taking by incorporating the "A" weighting which corresponds to the ear's actual frequency response

be seen, the relationship between loudness and frequency becomes more pronounced at very low listening levels. This is why manufacturers of stereo equipment include a *loudness* contour circuit in their amplifiers which selectively boosts the frequencies that the ear finds most difficult to hear at lower levels.

weighted scales: the "A" scale

To compensate for the ear's varying sensitivity to frequency, a weighted scale was created for measuring sound-pressure level. Termed the "A" scale, it follows the Fletcher-Munson curves and, as a result, de-emphasizes the middle frequencies of the audible spectrum while overemphasizing the high and low frequencies. As a result, this scale directly reflects the ear's perception of loudness. When the "A"-weighted scale is used, decibels are denoted by the term *dBA*. Thus, a 40-Hz tone of 77 dBA is just as loud as a 400-Hz tone of 77 dBA. This weighting system enables us to compare various signals without having to account for the ear's idiosyncrasies. Most sound-level meters are supplied with an "A"-weighted scale.

chapter two

how acoustics affect recording

introduction

Room acoustics can mean the difference between good and bad recordings. They affect every phase of the recording process, ranging from the initial recording session to the final cutting of the disc.

During a recording session room acoustics are omnipresent. They can cause instruments and voices to be clearly isolated and distinct or hopelessly muddled together in a common blur. They can create a natural ambience or a stilted aura of artificiality.

Acoustical qualities can also have an important secondary effect at the recording session—on the temperament of the performers. When the acoustics are excellent, the musicians are apt to be pleased, and will tend to perform more confidently and enthusiastically. On the other hand, when the acoustics are poor, the musicians are apt to feel incapable of communicating with one another and a strained quality or lack of cohesiveness may characterize the performance. Thus room acoustics can be metamorphosed into the emotional as well as the technical quality of a recording.

In small home studios, where music is often recorded and played back in the same room, room acoustics are especially important. In such cases, even a moderate problem can become exaggerated into a large acoustical dilemma. For instance, if a room were too reverberant, causing the music to sound "muddy" during recording, the sound will be further muddled during playback. With the original inaccuracy compounded, the resulting recording is apt to sound nothing like the original.

To eliminate composite problems such as this, critical attention must be paid to room acoustics in professional and home studios alike.

Thus the purpose of this chapter is to present the seven basic effects of room acoustics on recording and to demonstrate how these effects can be applied and controlled to produce better recordings.

noise

Noise, quite simply, is unwanted sound. It is the first and most basic element of room acoustics that affects recording. If a recording environment is noisy, chances are the recording itself will be.

Noise can be produced by an infinite variety of sources (from rush-hour traffic to screaming neighbors) whose only common bond is their undesirability. Noise can never be completely eliminated—for it is interwoven with the very fabric of our existence. Even the random motion of molecular particles makes noise. Our living bodies are virtual factories of noise. In fact, our own ears must make a degree of noise in order to function.

Despite these impediments, in a recording environment we attempt to reduce extraneous noise to its lowest possible threshold. In this book, the process is referred to as *"soundproofing."*

Ambient noise

Ambient noise may be thought of as the background noise always present in any environment. It can result from the din of traffic, the operation of air conditioners and other mechanical equipment, or much more subtle sources such as the rustling of trees or the patter of rain.

On a busy street, the ambient-noise level may be as high as 90 dBA (spl). A quiet farmhouse may have an ambient-noise level of only 40 dBA. *In a recording studio, experience has shown that the ambient noise-level must lie well below 25 dBA to be acceptable.* The above values, as well as other common ambient-noise levels, are shown in the accompanying table.

Some common misconceptions about ambient noise

There are several misconceptions about ambient noise and how it is controlled. The first such misconception is that bothersome ambient noise can be eliminated by applying carpet, curtains and acoustic tile to the surfaces of a room. The truth is that these materials do little or nothing to stop the bulk of sound energy from entering or leaving a room. They are too porous and lightweight.

Apparent Loudness	Source	Sound-Pressure Level
Painful	**Jet Aircraft** *at runway*	140 dBA
Extremely Loud	**Rock Band** *on stage*	130 dBA
	Subway Train *track side*	115 dBA
	Heavy Truck Noise *roadside*	90 dBA
	Noisy Office *many typewriters, office machines, etc.*	80 dBA
Loud	**Normal Traffic** *roadside*	70 dBA
	Average Office Noise	60 dBA
Moderate	**Moderate Speech**	55 dBA
	Tranquil Residential Neighborhood *outdoors, night*	45 dBA
Faint	**Residential Neighborhood** *indoors, night*	30 dBA
	Rustling Leaves	20 dBA
Very Faint	**Normal Breathing**	10 dBA
	Threshold of Audibility	0 dBA

Figure 2-1 Common noise levels

The chief effect of these materials is to change the acoustical character of sounds made *within* the room. By reducing the number of high-frequency reflections that a sound will undergo inside the room, carpet, curtains and acoustical tile make a room seem "dead." These materials do nothing to stop the penetration of *low-frequency sounds* (traffic noise, machinery sounds), which constitute the largest component of most ambient noise. Therefore, these materials are not the solution to the problem of lowering ambient noise. (The proper methods for controlling noise will be presented in the next chapter on soundproofing.)

Another common misconception is that a room that sounds quiet to the ear has a low ambient-noise level. This is false because of the human ear's remarkable propensity for rejecting background noise. This psychological phenomenon has been incorporated into the hearing system as a protection against sensory overload by the millions of sounds that surround us. As a result, we seldom even notice sounds which constantly envelop us, such as traffic, wind or continuous mechanical noise. Unfortunately, a microphone picks up these noises immediately and transfers them to the recording.

Another important reason that the microphone is more sensitive to ambient noise than a human stems from the fact that a mike is monaural (i.e., *one ear only*). Any sound which creates a positive pressure in the microphone diaphragm will create a corresponding impulse in the microphone. In contrast, the human hearing mechanism utilizes a *binaural* or two-ear system. This enables the brain to assess time lags and phase differences between sounds arriving at the left and right ears. As a result, it can pinpoint a sound it "wants" to hear while simultaneously *suppressing* the more omnidirectional background noise. A microphone does not possess this ability and will therefore register much more background noise than a human ear. Since it is a microphone that records sound in the studio, simple listening tests to determine ambient-noise levels are not adequate.

Measuring ambient noise

To determine the actual level of ambient noise in a room, readings must be taken with a *microphone* and a *sound-level meter.* Readings are taken several times during the day (ambient noise changes throughout the day) at various frequency bands across the audible spectrum. The maximum results are plotted to produce an *ambient-noise curve.*

The height and shape of the curve indicates the extent of the noise problem in a particular room. To determine exactly how much soundproofing will be required to reduce ambient noise to an acceptable level for recording, the noise curve is plotted on the same graph with a curve showing acceptable noise levels for a recording studio. The difference, in dB, between the two curves shows how much soundproofing will have to be done.

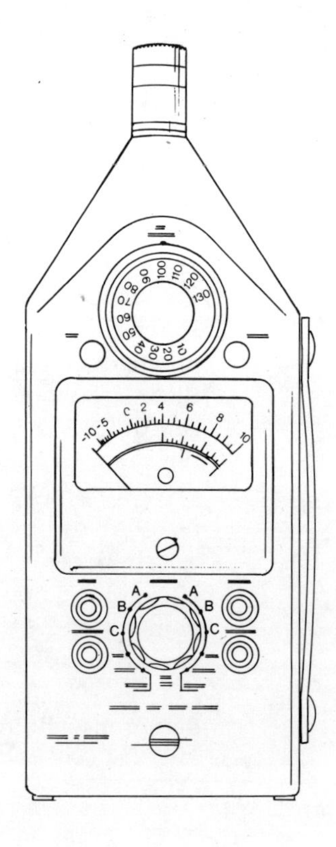

Figure 2-2 A typical sound level meter (courtesy of B&K Instruments)

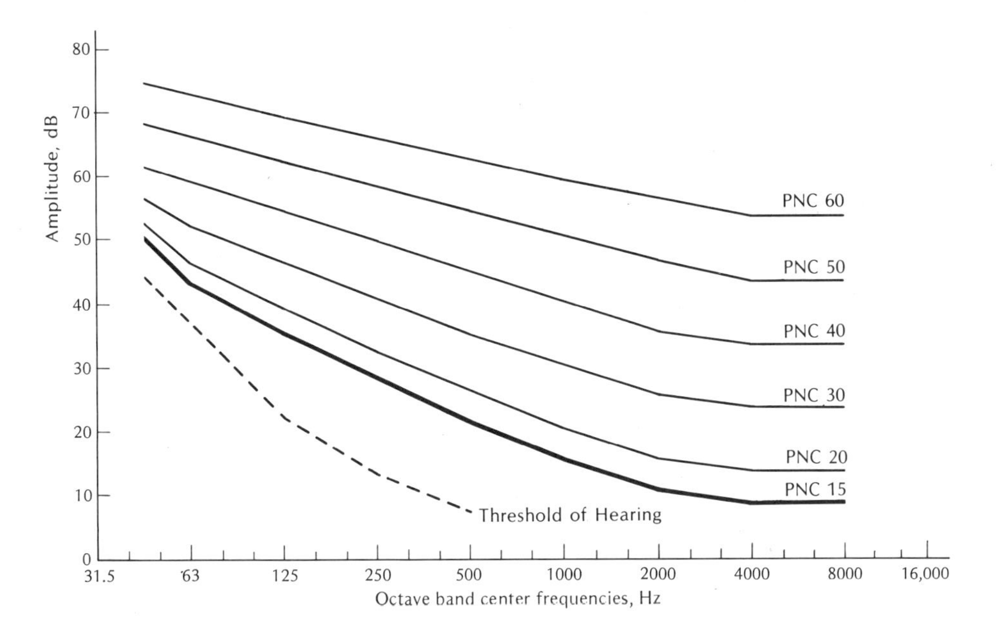

Figure 2-3 Maximum ambient noise level for a good recording studio: PNC 15

PNC (Preferred Noise Criteria)

Ambient noise is often specified in terms of a single number rating system called the *PNC (Preferred Noise Criteria) system.*

Test results in a specific location are compared to a group of standard PNC curves and the standard curve which best approximates the test results, according to a detailed set of rules, is chosen to represent the noise level.

PNC curves slope upward in the low frequency range in order to simulate the ear's reduced sensitivity in this portion of the spectrum.

The maximum ambient noise level for a good recording studio is typically *PNC 15,* which is shown in the accompanying illustration.

Determining the required soundproofing

After the maximum noise spectrum outside the studio site has been measured during representative times of the day, to include such intermittent occurrences as airplane flyover and truck passby, the results are plotted on a graph along with the maximum acceptable noise inside the studio (usually *PNC 15*).

The difference between the two curves at each octave band represents the soundproofing required.

For example, looking at Figure 2-4, if the difference between the two curves is 52 dB at 63 Hz, 61 dB at 500 Hz and 60 dB at 8,000 Hz, the studio construction will have to provide at least this much *attenuation* in order to provide adequate noise protection. The amount of attenuation provided by a wall, ceiling or floor is referred to as its *Transmission Loss (TL)* and is measured in *dB*.

Sometimes it is more important to stop sounds from *leaving* the studio than it is to stop sounds from entering. This situation might occur when a studio is being planned in a quiet residential area where neighbors are all too eager to complain about the faintest hint of loud music.

In this case, the existing outside ambient noise level is plotted as a *reference noise floor*. Then, the noise level expected *inside* the studio is plotted on the same graph (see Figure 2-5). The numerical difference between the two curves at each octave band represents the soundproofing needed to reduce sounds in the studio to a point where they will generally not be noticed outdoors.

To determine which requires more soundproofing — noise coming in or noise going out, the results of both upper and lower graphs are compared at each octave band. The greatest differential at each band is always selected as the final *soundproofing guideline*.

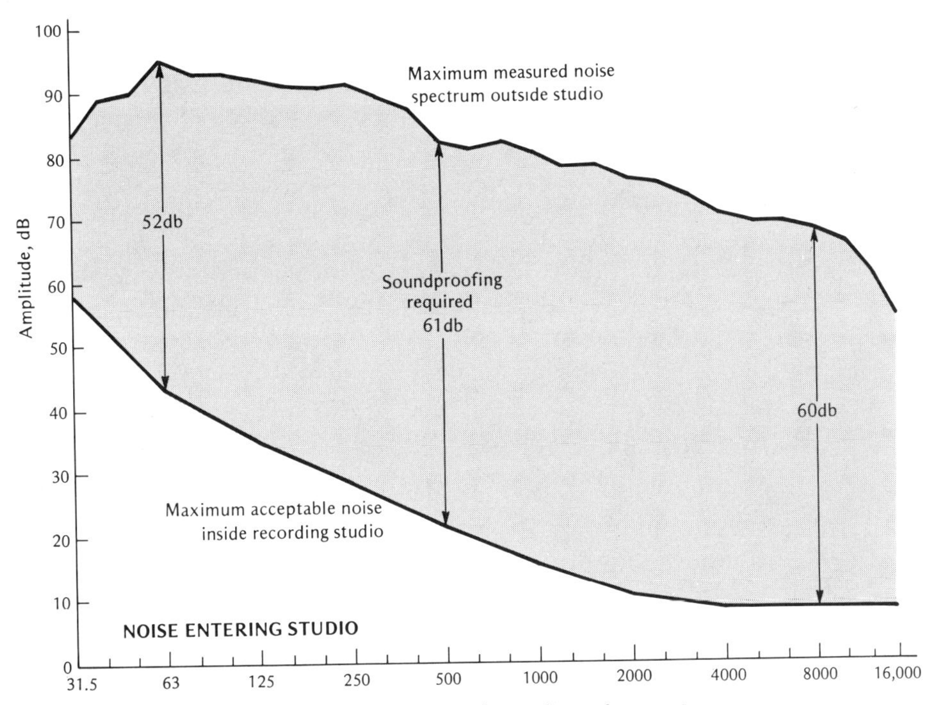

Figure 2-4 Determining the soundproofing for noise entering the studio

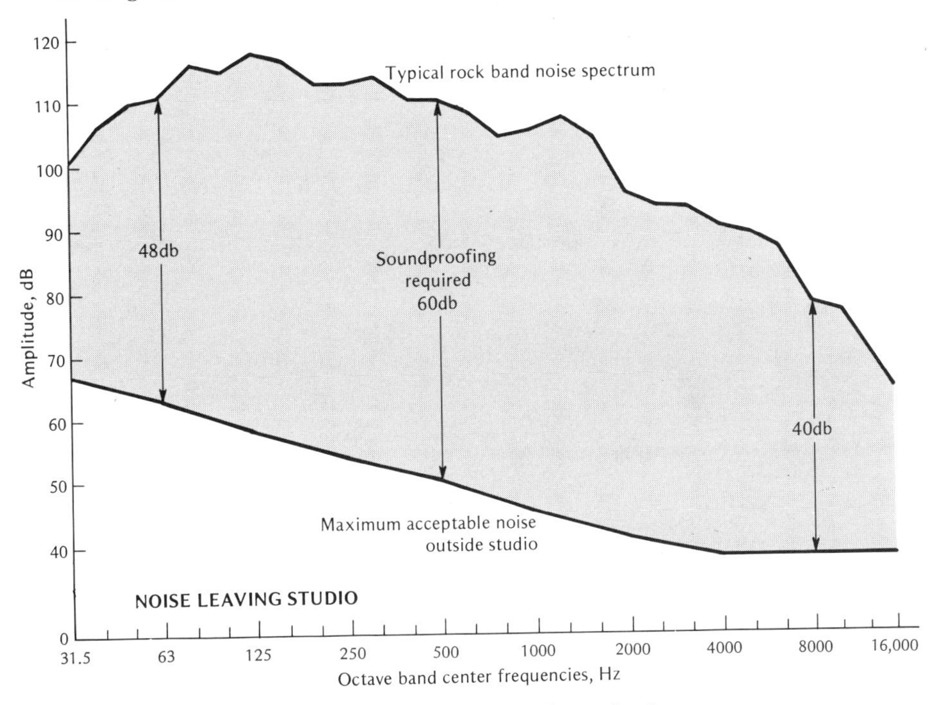

Figure 2-5 Determining the soundproofing for noise leaving the studio

reverberation

Reverberation is the gradual decay of sound in a room, after the source has ceased. It affects the character of all sounds in a room and, hence, the character of any recording being made. Compare the character of your voice in a closet with the character of your voice in a bathroom. In the closet, all of the clothing helps to *absorb* the sound of your voice. There is little reverberation. Hence, your singing sounds "weak" or "dead." In a bathroom, on the other hand, there is a great deal of reverberation. The reflective surfaces of tile and porcelain *reflect* the sound of your voice back to you. These reflections reinforce your voice and made it sound powerful, resonant and "live."

In the studio, we want to be able to control the amount of reverberation and adjust it to individual instruments and voices to suit our taste. Too much reverberation makes music "muddy" or unintelligible. Too little reverberation makes music "dry" or lifeless.

How is reverberation formed?

When sound waves leave a source, they travel outward in a three-dimensional arc. Some of these waves travel directly to the listener and are called *direct waves*. The other waves (the greater proportion) strike the walls,

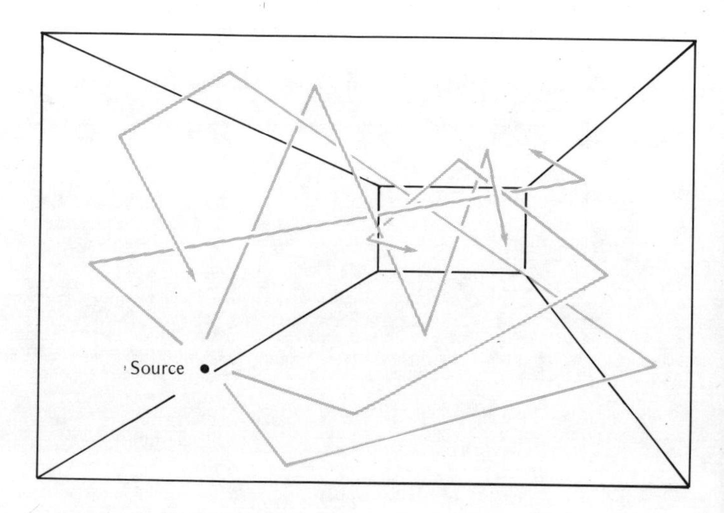

Figure 2-6 Reverberation: A multitude of decaying reflections

floor, ceiling and room furnishings and are reflected back into the room. These *reflected waves* eventually strike other surfaces and the process repeats. Since the waves lose energy during travel, as well as each time they strike a surface, successive reflections become gradually weaker until they eventually reach inaudibility.

These reflections arrive at the listener in a continuous stream. They are so closely spaced that the ear is incapable of distinguishing them as individual sound waves. Instead, the ear perceives a gradual decay of the room sound, sometimes lasting several seconds after the source has stopped altogether.

The time taken for the sound to decay to one-millionth of its original intensity (i.e., 60 dB of decay) is called the *reverberation time* (commonly called *reverb time*) and denoted by the letter "T."

The reverberation time is a function of two main factors:

- the volume of a room
- the absorptivity of a room

In a large room, a wave must travel long distances before reaching a reflecting surface. Therefore, *the rever-*

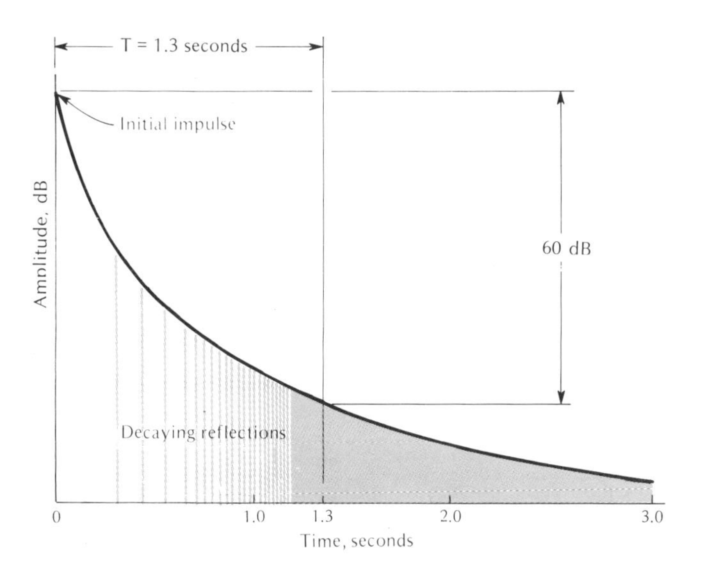

Figure 2-7 The calculation of reverberation time (T)

beration time is generally longer for large rooms. The reverb time for a huge cathedral may be 6 or 7 seconds. For a small living room, the reverb time may be only half of a second.

If the surfaces of the room are very absorptive (with carpet, drapes, thickly upholstered furniture), very little wave energy will be reflected back into the room at all. Thus, absorptive rooms generally have shorter reverberation times.

Computing reverberation time

Reverberation time, room size and absorptivity are related according to the following basic formula, developed by W.C. Sabine:

$$T_{60} = \frac{.049\,V}{A}$$

where:

T_{60} = reverberation time (60 dB of decay), seconds

V = room volume (length × width × height), cu. ft.

A = total absorptivity of room, measured in Sabins

Figure 2-8 Actual Decay curve for a room 12′ × 17′ × 8′

Because the formula does not take into account such variables as frequency of source, room geometry, temperature and humidity of air, and presence of people, furniture and other random obstructions that increase the scattering of sound waves, the formula provides only an approximation.

Other formulas

Other formulas, which take into account some of the variables mentioned above, have been developed to calculate reverberation. The most commonly used of these is the Carl F. Eyring Formula listed below.

$$T_{60} = \frac{0.049\ V}{-S\ \ln\ (1 - \bar{a})}$$

where:

T_{60} = reverberation time, seconds
V = volume of room, cu ft
l_n = $\log_e$
$\bar{a}$ = average absorption coefficient of the room surfaces
S = total surface area of room, sq. ft.

Measuring reverberation

Since many random factors often affect reverb time in ways that cannot be predicted by the formulas, reverberation is often measured rather than computed for an actual environment. Although several methods for measuring reverb time exist, all involve a *source* (such as pink noise), a *filtering system* (to enable testing at various frequencies), a *microphone* and an instrument to accurately register the decay, such as a *graph plotter* or a *digital counter.* Several measurements are taken at various room positions, and the results are usually averaged to give an overall representation.

Live vs. dead rooms

Rooms which have hard surfaces (tile, wood, plaster) and long reverberation times (over one second) are usually called *live* rooms These types of rooms have the net effect of reinforcing the sound of music played in them (especially at lower frequencies) and causing the music to sound relatively loud and resonant within the room.

Many musicians are quite pleased to hear themselves loudly reinforced and prefer playing in rooms that are slightly live. Reverberation also tends to smooth out the inconsistencies in timbre and pitch, thus making a given performance sound more consistent and professional. Perhaps this explains why singing in the shower is so popular. The hard-tiled surfaces of the shower reflect the voice, emphasizing its lower-frequency overtones, while smoothing inconsistencies of pitch and timbre—giving the illusion of tremendous depth, power and control.

Rooms which have soft surfaces (curtains, carpet) and short reverb times (under .5 seconds), are called *dead* rooms. They give little or no reinforcement to music played within, causing it to sound weak and lifeless. These types of rooms make small inconsistencies in timbre and pitch readily apparent to a listener (and a musician) and, as a result, are quite unpopular among musicians.

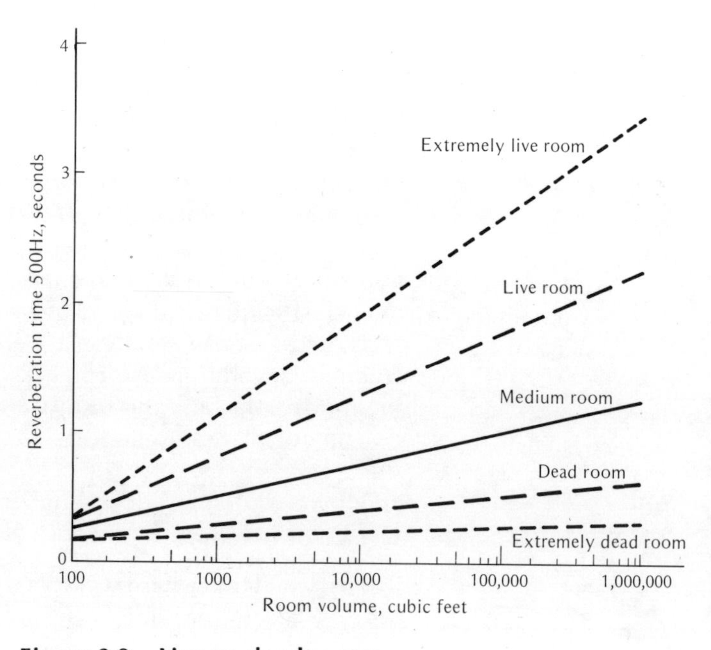

Figure 2-9 Live vs. dead rooms

What is the optimal reverberation time?

The fact that reverberation can help a voice or instrument to sound fuller and more "professional" might be taken to mean that a recording studio should have a great deal of reverberation. However, this is not true. Determining the optimum reverberation time is a complicated matter.

In a multitrack rock studio, for example, instruments are *isolated* from one another and routed to separate channels of the multitrack tape for mixing and balancing at a later time. Too much reverberation in the studio would make isolation almost impossible to achieve; reflections from various instruments and amplifiers would bounce aimlessly, finding their way into microphones meant for other instruments. This effect is especially pronounced at lower frequencies, whose wavelengths are quite long and difficult to absorb. In the extreme case, this misdirection of sound, called *leakage,* would cause many instruments to appear on each track of the multitrack tape, thus defeating the entire purpose of multitrack recording. It would mean that when the level of a particular instrument was to be raised or lowered, the sound of other instruments would be inadvertently raised and lowered as well. The overall recording would have a rather loose or muddled quality, reminiscent of recordings made with a single microphone in a basement or garage. Thus, for a studio specializing in multitrack rock music, a long reverberation time would be intolerable.

Another important reason that excessive reverberation is discouraged in multitrack studios is that electronic means are readily available for adding controlled amounts of reverberation to a recorded sound *after* the initial taping has taken place. Thus, if a certain amount of reverberation is desired, the voice or instrument can be taped without any reverberation whatsoever, and the correct amount can be *added* electronically at a later time. Since any reverberation present on the recording cannot be *subtracted,* it is a much safer technique to record an instrument slightly *dry* (no reverb) at first and add the desired amount of reverberation later, during mixdown.

A final reason that overly reverberant rooms do not

make good general-purpose studios stems from the fact that a listening environment always adds a little of its own reverberation to the material on a recording. Therefore, when heard over loudspeakers, all recordings will seem to be more reverberant than they actually are. This effect is especially pronounced on recordings with short reverb times listened to in large reflective rooms. However, since the average living room has a reverb time of only .6 seconds, the effect is only moderately detectable in most home listening situations. This reverberation summing effect does *not* occur when listening through headphones, thus explaining why a record heard through phones always sounds crisper and more distinct than the same record heard over loudspeakers.

These arguments against excessive reverberation all seem to suggest that a recording studio should have a

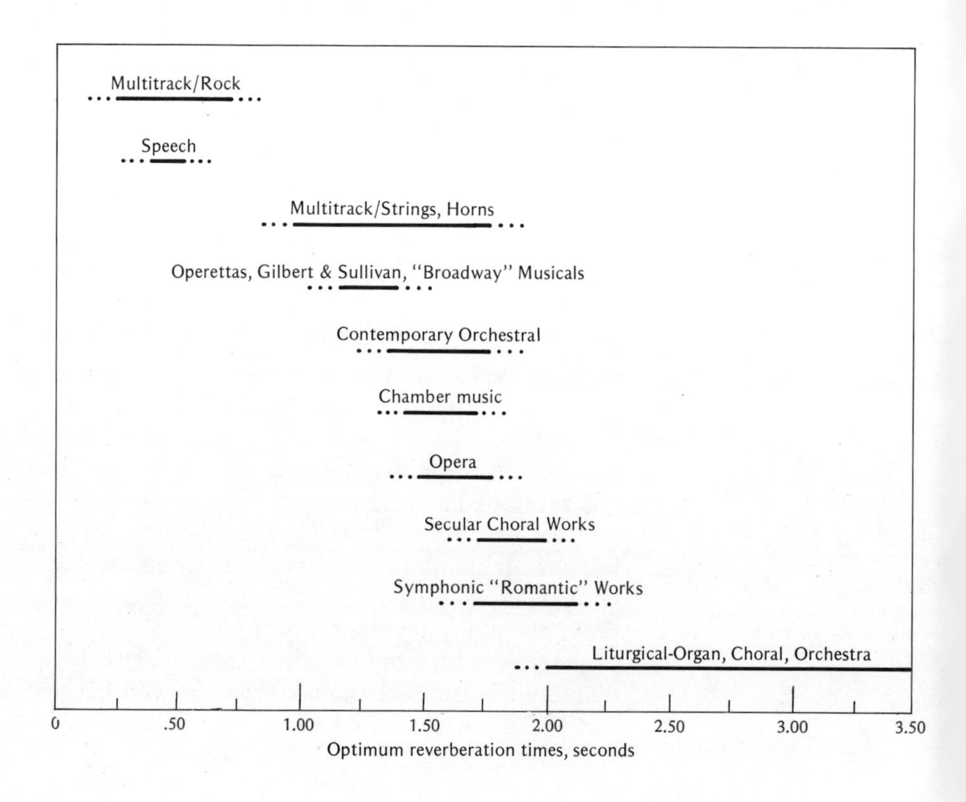

Figure 2-10 Optimum reverberation times (500-1000 Hz)

short reverb time. However, if the reverb time is too short, the studio will become very anechoic or dead. The music will sound dull and lifeless, and musicians will feel uncomfortable. *So what is the optimal reverb time?*

The answer is that optimal reverb time will depend entirely on the type of music to be played. In a large studio for classical music, a certain amount of reverb will not only be tolerated, it will be desired. This will give the orchestra a *group sound,* blending inconsistencies between various players and sections. Classical studios may have reverb times of several seconds. In a rock studio, where levels are loud, space is tight and isolation is critical, leakage must be avoided and reverb time must be fairly short (in the .25 to 1.0-second range).

If a studio is to accommodate as many types of music as possible, we must look into ways of making reverb time *adjustable.*

In general, reverberation must be reduced at low frequencies—the worst culprits for leakage—and adjusted at mid and high frequencies to suit the music played.

absorption

When a sound wave strikes a surface, three phenomena occur:

1. A portion of the sound energy is *transmitted* through the barrier.
2. A portion of the sound energy is *reflected* back into the room.
3. A portion of the sound energy is *absorbed.*

When we say that sound has been "absorbed," we simply mean that it has been converted into another form of energy, generally heat. Admittedly, the energy levels normally encountered in sound waves are so minute (a billionth of a watt per square centimeter) that the heat of absorption is scarcely detectable. However, this conversion phenomenon occurs to satisfy the laws of physics.

The fraction of incident sound energy absorbed by a material is called the *absorption co-efficient* or absorptivity. Therefore, if a certain material absorbed 83% of all incident sound energy at a particular frequency, its absorption co-efficient would be .83 at that frequency.

The absorption co-efficient of a material depends on many internal characteristics (including porosity, modulus of elasticity, internal-flow resistance and thickness). It also depends on external factors, such as the frequency of the incident wave, the angle of the incident wave (absorptivity is greatest when a wave strikes perpendicular to the absorber) and the manner in which the absorber is mounted.

Porous absorbers

Almost without exception, conventional acoustic absorbers, such as acoustical tile, carpeting, curtains, mineral wool and fiberglas are *porous absorbers*. Measuring only several inches or less in thickness, they absorb more effectively at the high frequencies. This results directly from the fact that high frequencies have small wavelengths (from several inches down to only fractions of an inch long). The to-and-fro motion of these small waves gets easily trapped in the tiny air spaces of the porous material. The resulting friction converts the sound energy into heat and the wave energy gets absorbed. Low

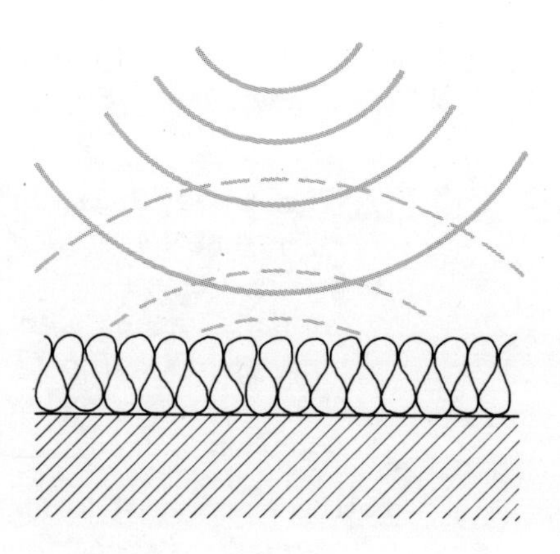

Figure 2-11 A porous absorber

frequencies, with large wavelengths (up to 50 feet long) do not get trapped in these tiny spaces and, as a result, are not absorbed.

Membrane absorbers

The method of porous absorption is not the only means of acoustical absorption. A second type of absorber is the flexible panel or *membrane absorber.* This consists simply of a membrane, such as plywood, mounted over an air space of several inches or more. A membrane absorber works in the following way: An approaching sound wave strikes the cover panel, setting it into motion. The panel resonates at a frequency determined by the stiffness of the panel and the size of the air space behind it. The resonating panel dampens the approaching sound waves in the same frequency range of the panel resonance. Thus, these particular frequencies are absorbed, while other frequencies are reflected. Membrane absorbers are absorptive at low frequencies, in contrast to porous absorbers, which are absorptive at high frequencies. Thus, they are often called *basstraps*.

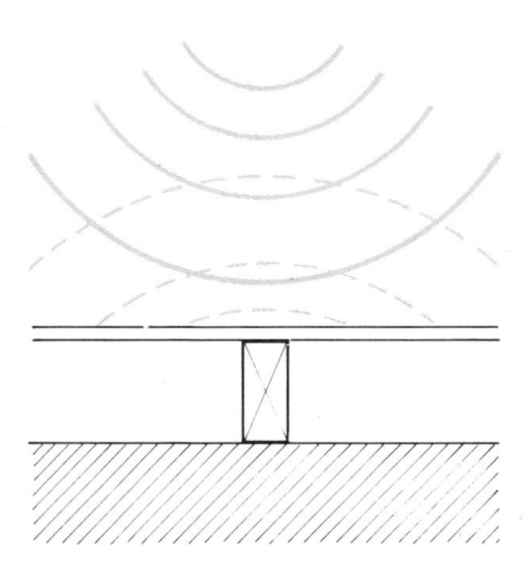

Figure 2-12 A membrane absorber

Helmholtz resonators

Another method of absorption utilizes a device called a *Helmholtz resonator*, which consists of an enclosed air space with a small opening or *neck*.

A gallon jug is a good example of a Helmholtz resonator. As is commonly known, a gallon jug will produce a pitched tone when air is blown into the neck. As the jug is filled with water, the enclosed air volume gets smaller and the pitch of the tone rises. This phenomenon occurs because each such resonator has a natural or resonant frequency, determined by the volume of air enclosed and the size of the neck.

This principle can be used to *absorb* sound in the following way: When a sound wave at or close to the natural frequency of the resonator strikes the opening, the air in the *neck* is set into vibration. The enclosed volume of air resists compression and the oscillations in the neck are sent back out towards the approaching wave. Phase cancellation occurs and the approaching wave is absorbed.

This type of resonator can be designed to produce ab-

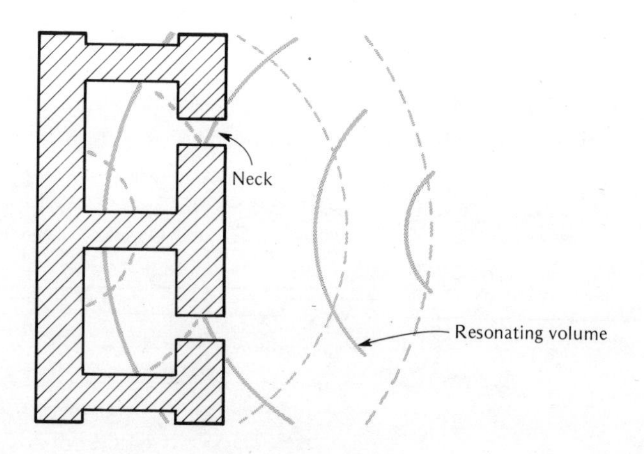

Figure 2-13 A Helmholtz absorber

sorption at very specific frequencies. Thus, its principal use in the studio is to eliminate severe room resonances and standing waves, especially in the lower-middle frequency range.

Rather than using gallon jugs, studios commonly employ concrete or wooden cavities in the wall to provide the air spaces and punched holes to form the interconnecting necks. Very often, "multiple" necks (i.e., many punched holes—as may be found in a perforated sheet of pegboard) are used to increase the usable surface area of the absorber.

Broadband absorbers

Another method for absorbing sound is called the *broadband absorber*. This method utilizes a series of *hanging baffles* to obstruct the travel of the sound waves. These baffles are constructed of porous materials, such as fiberglas (sometimes with a stiffer inner structure). When an approaching sound wave strikes the baffles, it is caused to diffract, thereby hitting other baffles in the immediate vicinity. The process repeats, eventually dissipating the sound energy of the wave.

In order for the broadband absorber system to work properly, the baffles must be hung in predetermined configurations. Each baffle must be spaced at a critical distance from the baffles surrounding it. The configuration and critical distance are determined mathematically and vary from one application to another. In general, the baffles must be spaced far enough apart to allow the sound waves of the appropriate frequency to enter, yet close enough to trap and absorb the energy once it enters.

Since the effectiveness of this type of system at low frequencies is a function of the size of the baffle itself, the dimensions of the baffle configuration can be adjusted or "tuned" to a desired cut-off frequency.

The baffle system is sometimes inappropriately called *basstrapping*. In actuality, the baffle system is a broadband absorber with a tunable, low-frequency cut-off.

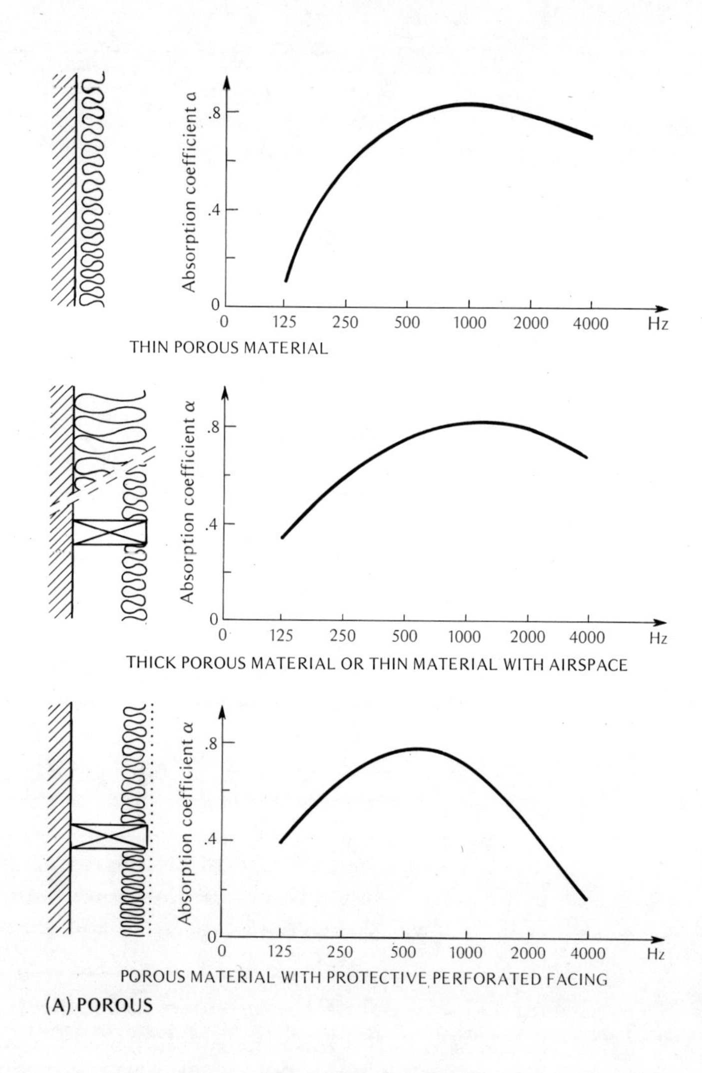

Figure 2-14 Sound Absorption: Summary

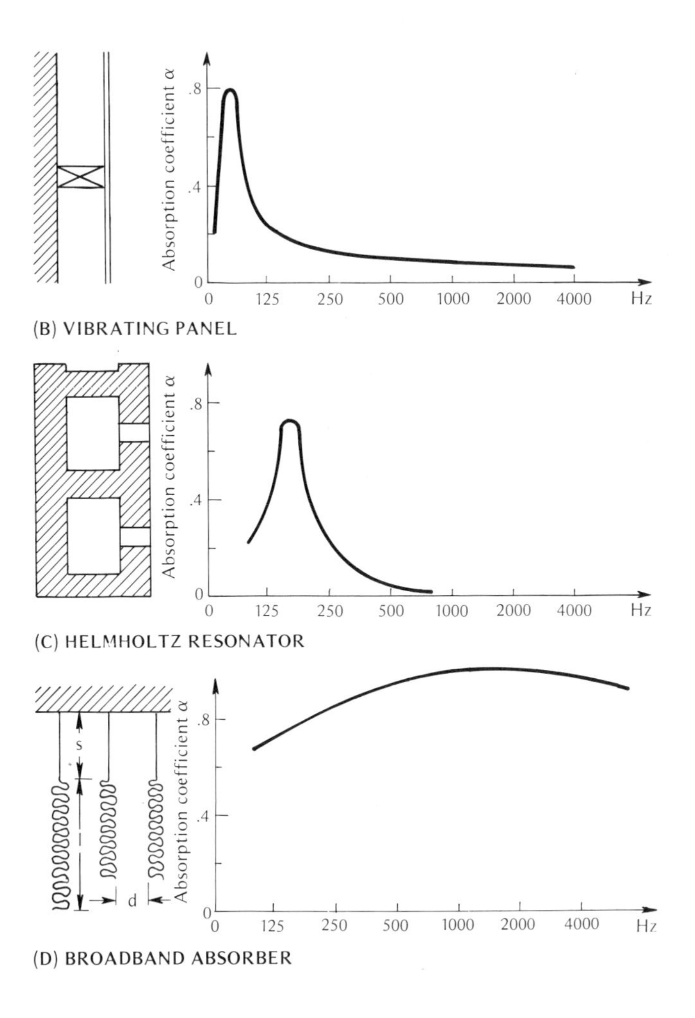
Absorption coefficient α
.8
.4
0
0
125
250
500
1000
2000
4000
Hz
(B) VIBRATING PANEL
Absorption coefficient α
.8
.4
0
0
125
250
500
1000
2000
4000
Hz
(C) HELMHOLTZ RESONATOR
s
d
Absorption coefficient α
.8
.4
0
0
125
250
500
1000
2000
4000
Hz
(D) BROADBAND ABSORBER

coloration

Sometimes, as a result of its size and geometry, a room seems to "prefer" to accommodate the reverberation of certain frequencies. This results in the reinforcing and lingering of certain tones when music is played—after they should normally have ceased. This effect is called coloration. It amounts to an alteration of the natural sound.

Coloration is a result of standing waves or room resonances. These are waves whose original vibrations are continuously reinforced by their own reflections. A typical room has many standing waves and potential colorations. In the recording environment, these must all be eliminated.

How do standing waves form?

Standing waves can form in several ways. In the simplest case, a low-frequency sound wave will resonate between two opposing surfaces of a room, continually reinforcing its amplitude by the method of constructive interference (i.e., each reflection adds to the previous one).

This type of standing wave is called an *axial mode* and will occur at frequencies whose wavelengths are twice as long as the distance between the reflecting surfaces. Thus, *the lowest standing wave that occurs in a room has a wavelength twice as long as the room's largest dimension.*

Axial standing waves can occur in *each* dimension of a rectangular room (length, width and height) and can be calculated using the following formula:

$$f_o = \frac{V}{2\,d}$$

where:

f_o = fundamental frequency of the standing wave
V = velocity of sound (1130 ft. per second)
d = room dimension being considered, ft. (length, width or height)

Other standing waves will also form at the *harmonics* of the axial frequencies.

Thus, for a room whose length, width and height are 20, 16 and 8 feet, respectively, simple standing waves would form at the following frequencies:

	Harmonics (Hz) 1st	2nd	3rd	4th
$f(1) = \frac{v}{2(d)} = \frac{1130}{2(20)} =$	28	56	84	112
$f(w) = \frac{v}{2(d)} = \frac{1130}{2(16)} =$	35	70	105	140
$f(h) = \frac{v}{2(d)} = \frac{1130}{2(8)} =$	70	140	210	280

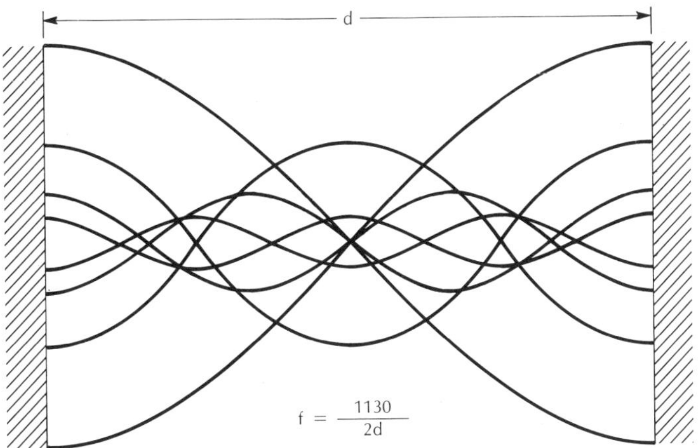

Figure 2-15 The formation of standing waves

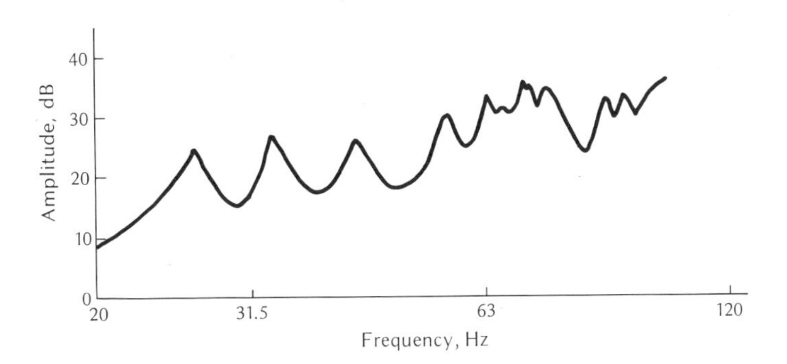

Figure 2-16 The presence of standing waves in a typical room

The net effect of the presence of these standing waves would be a selective boosting or reinforcement of the room sound at these particular frequencies, resulting in a tremendous imbalance or coloration of the acoustical properties of the room.

Other types of standing waves, known as *tangential* and *oblique modes,* also exist. Numbering in the hundreds, these modes have complex mathematical derivations. Fortunately, by the time a studio is filled with people, equipment, instruments and baffles, the path of sound is obstructed and most of these latter modes are eliminated. As a result, they play only a small part in determining the acoustical climate of the studio and will not be dealt with in detail.

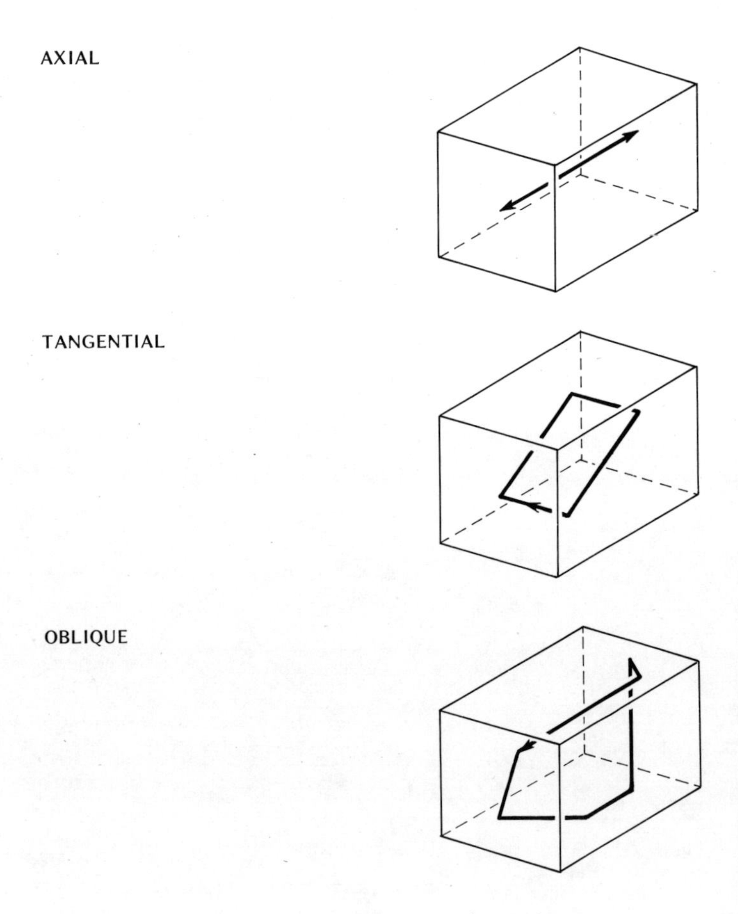

Figure 2-17 Three principle types of standing waves

Coloration in the studio and control room

Most rooms have dozens of severe standing waves affecting their acoustical properties. In the *studio,* the presence of these characteristic colorations poses two main problems:

1. The natural sound of the instruments being recorded is altered.
2. The reverberation time at the standing wave frequencies (or at low frequencies, in general) is exaggerated, causing excessive *leakage* between instruments.

In the *control room,* the presence of these standing waves causes the following problems:

1. The perceived frequency balance of the recording is altered, making it impossible to determine what balance is really on the tape and what is being manufactured by the room acoustics. (Is the bass as boomy as it sounds, or is it the room?) This situation can cause engineers and producers to overcompensate or undercompensate with level and equalization, sometimes ruining a recording.
2. Recordings mixed in the control room will sound quite different when played back in another room, even when the same type of speakers are used.
3. The reverberation time at the standing wave frequencies is lengthened, causing a lack of clarity at these frequencies (i.e., "muddy" sound, especially in the bass). This is by far the most common problem in listening and mixing rooms.

The remedy for these problems includes the application of absorbers specifically designed to eliminate colorations. (This is discussed in Chapter IV: *The Studio*.)

echo

Echo is the repetition of a sound due to the delayed arrival of a reflected sound wave. In order for a reflected wave to be interpreted as a distinct echo, it must arrive at the ear more than 35 milliseconds after the initial impulse. Sounds arriving before this time are indistinguishable from the original and are heard as reinforcements of the original wave.

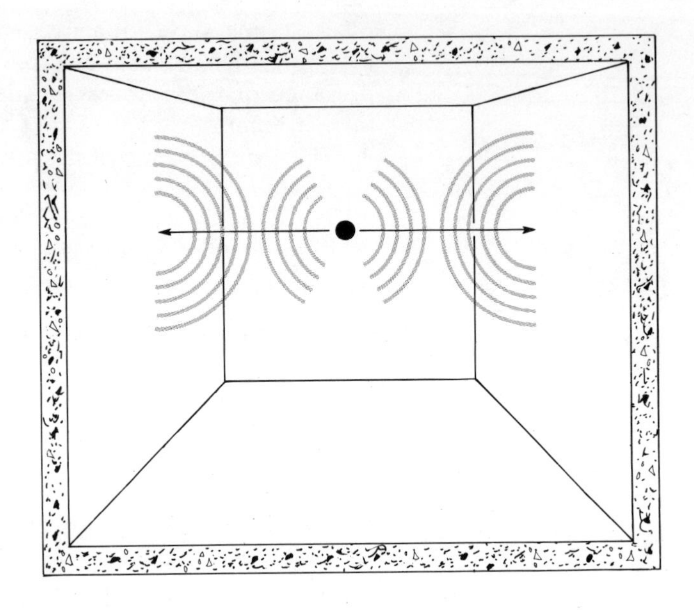

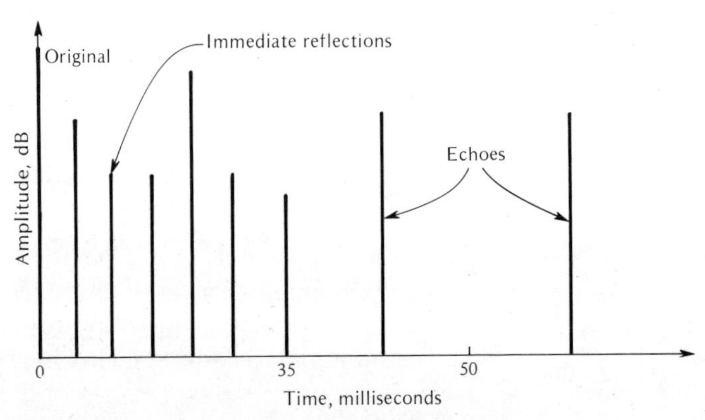

Figure 2-18 Echo: The repetition of sound

Since the time discrepancy between the arrival of a direct and reflected sound wave depends on the distance to the nearest reflecting surface, large rooms (over 3,000 cubic feet) are more likely to have distinct echoes than small rooms. Small rooms, with hard, opposing parallel surfaces, are likely to have a short, continuous stream of echoes. These are called *flutter echoes.*

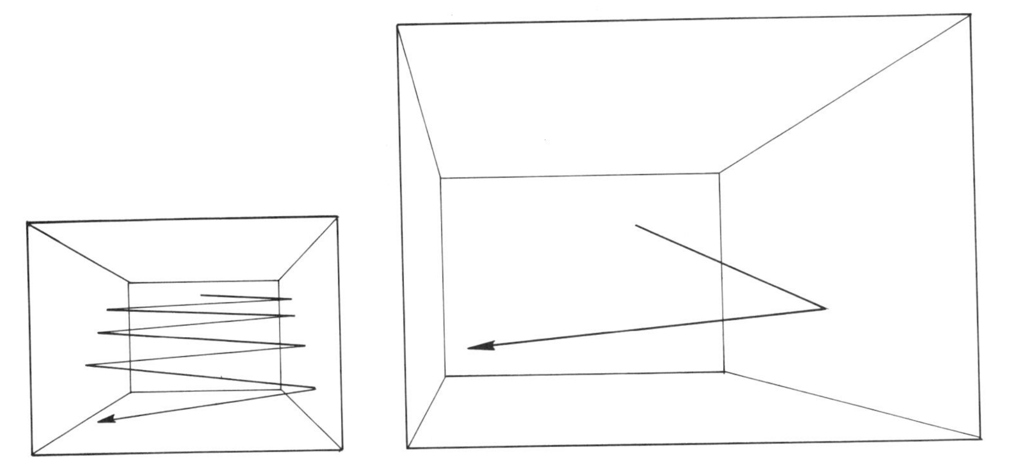

Figure 2-19 Echoes in large vs. small rooms

Echo in the studio

Echoes in the studio can completely ruin a recording. They are especially troublesome with percussion instruments that have high initial amplitude peaks and fast decay times. The high initial amplitude enables the echo to be high in energy, while the fast decay enables the echo to be easily distinguished from the original. As a result, echoes interfere substantially with piano, bass and other percussive rhythm instruments.

The most common cause of echoes in the studio is the hard, reflective surface of the control room window. Oftentimes, a stream of *flutter echoes* will set up between this surface and other surfaces in the studio. Occasionally, only a single reflection will bounce off the window and return to the microphone in the recording area. This latter type of echo is called *slapback echo.* Both types must be eliminated.

Echo in the control room

Control rooms are not usually large enough to suffer from distinct delayed echoes. However, many control rooms are plagued by even a worse enemy—flutter echo. This stream of echoes often sets up between the control room glass and the other reflective surfaces of the room. Occasionally, the hard metallic surfaces of the

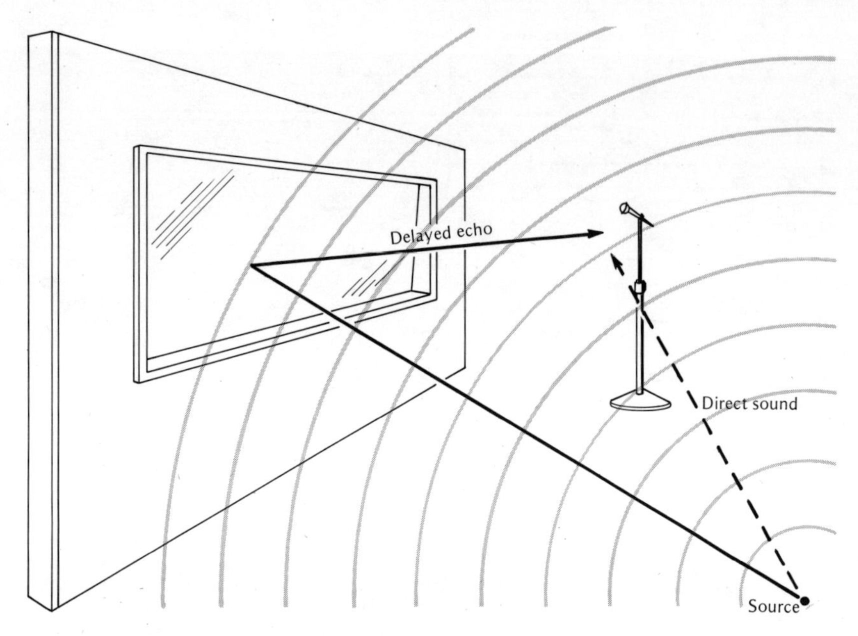

Figure 2-20 Echo caused by the control room glass

tape machines themselves will induce a particular type of flutter echo. Needless to say, flutter echo in the control room severely distorts a mixer's perception of what is on tape and must be eliminated completely.

diffusion

Diffusion refers to the homogeneity of sound in a room. A sound field is diffuse if its intensity at all points in a room is approximately equal.

Outdoors, it would be impossible for a sound field to be truly diffuse. The intensity of the field would always be greatest near the source and would always decrease rapidly as the distance from the source increased. Indoors, it is possible for a sound field to be scattered into thousands of reflections, thus enabling the sound (energy) to become relatively constant throughout the room.

The ability of a sound field to be diffuse indoors is a function of the room geometry. If the room is composed of concave geometries (i.e., a domed ceiling), diffusion

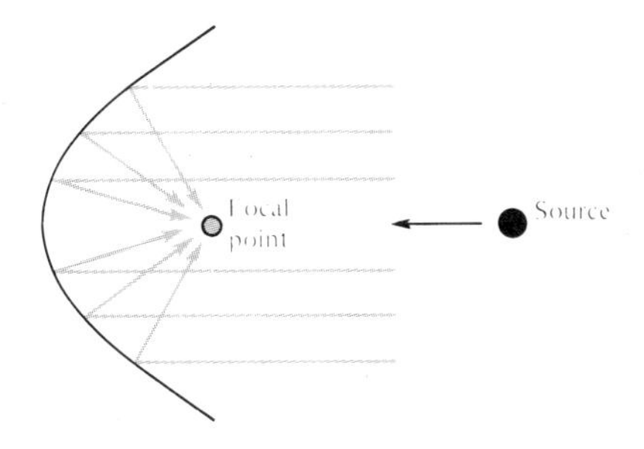

Figure 2-21 Focusing caused by a concave geometry

would be impossible. The dome would focus sound energy to a single location, leaving large discrepancies in amplitude from one place to another within the room.

If a room is composed of hard, flat, parallel surfaces (as most rooms are), diffusion will be only slightly better, owing to back-and-forth reflections between opposing parallel walls.

However, if a room is composed of extremely irregular surfaces and has many furnishings and other objects which bounce incident sound waves in random directions, diffusion will be greatly increased.

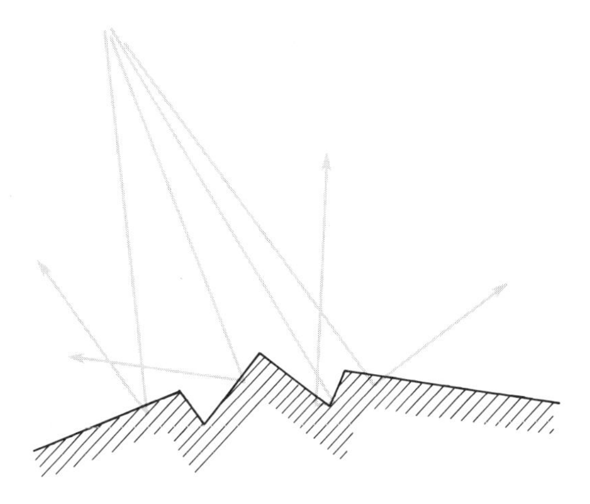

Figure 2-22 Sound waves scatter from an irregular surface causing diffusion

The ability of an irregular surface to diffuse an incident sound wave depends on the particular geometry of the surface and the frequency of the sound wave. In general, larger irregularities diffuse lower frequencies, and smaller irregularities diffuse higher frequencies. In order to have a measurable effect on diffusion, the surface irregularities must measure at least one-seventh of the wavelength being diffused. Therefore, in order to diffuse a sound wave of 50 Hz (wavelength = 22.6 feet), wall diffusers would have to protrude more than three feet from the wall surface.

The direct and reverberant fields

Because of the phenomenon of diffusion, most sound fields in rooms can be divided into two areas: the *direct field* and the *reverberant field.* The *direct field* refers to the area in the immediate vicinity of the sound source, where the fluctuations of the field are a direct result of changes in the source. The direct field tends to be erratic and unpredictable in nature. The *reverberant field,* on the other hand, refers to the areas removed from the source, where the many reflections (of the room) have created a reverberant condition that is uniform and diffuse.

Sound-level readings in the direct field would fluctuate wildly due to the close proximity of the source. On the other hand, sound-level readings in the reverberant field

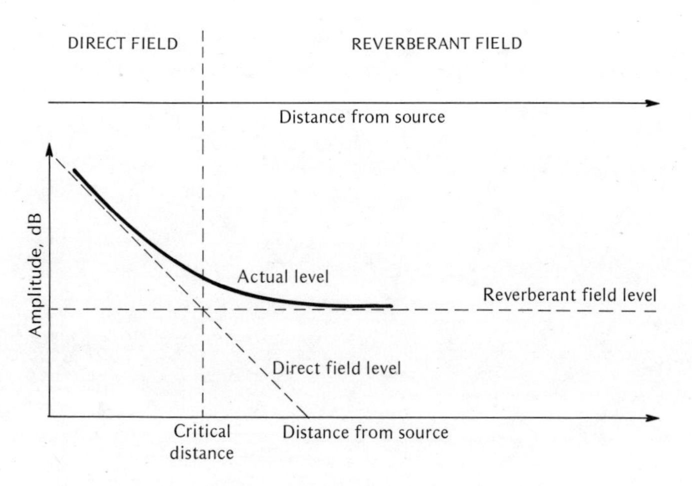

Figure 2-23 Direct and reverberant fields

would be more stable as a result of the diffuse acoustical condition.

The boundary between the direct and indirect fields occurs at the point where the direct energy equals the reverberant energy. This is called the *critical distance.* In a room 15 feet by 20 feet, the critical distance might occur at about 7 feet from the source.

Diffusion in the studio

When studios were used primarily for orchestral recordings, diffusion was sought after because it helped produce well-blended recordings. The reverberant field or *ambience* of the studio was the most important part of the recording. The advent of multitrack recording and high-level rock music has changed the role of diffusion in the studio. Now, studios for rock music are more concerned with isolating the instruments than they are with blending them. Blending causes leakage, thus defeating the purpose of multitrack recording.

Studios designed for rock music should basically *contain* the sounds of individual instruments. This does not necessarily mean that the studio should be dead or anechoic. Nor does it mean that the studio should be compartmentalized into a series of booths for the various instruments. It simply means that the studio design should not go to great lengths to send reflected waves scurrying all over the room, as some older theories of acoustics might suggest.

Studios for classical music or string and horn music, on the other hand, require a great deal of diffusion to maintain acoustical consistency in their sound fields and to create the proper ambience.

Despite its changing role, the principle of diffusion still is important in studios; however, it must be handled carefully on a studio-by-studio basis.

Diffusion in the control room

In the control room, it is important that the sound of the monitor speakers be assessed accurately from various room positions—not just a single location. This implies a certain amount of diffusion in the control room. With the proper amount of diffusion the engineer can move freely to adjust equipment without sacrificing the accuracy of his aural judgment. This diffuse condition also enables others (such as producers and musicians) to hear the identical balance as the engineer so that judgments of the sound can be coordinated.

Unfortunately, most monitor speakers are themselves very *directional* in performance, thus making such a diffuse condition difficult to achieve. This means that the design of the control room must play a very crucial role in providing the diffusion necessary to create a large, accurate listening area.

The principle of diffusion in the control room is also used to control standing waves and echoes, both of which can ruin a listening environment.

summary

We have discussed briefly the basic ways in which room acoustics affect recording. These are:
- noise
- reverberation
- absorption
- echo
- coloration
- diffusion

Each must be controlled properly to produce a good recording environment. The methods of controlling noise are presented in Chapter Three: *How To Soundproof A Room*. The methods for controlling the rest are presented in Chapter Four: *The Studio*.

chapter three

how to soundproof a room

introduction

We have already established that noise is undesirable in a room where serious recording or listening is being done. In some cases, eliminating the extraneous noise is a simple task. For instance, if the noise emanates from appliances or electric fixtures such as air conditioners and fluorescent lights, the problem can be easily remedied by turning off the appliances or fixtures during recording. If, however, the noise emanates from outside sources, the only way to eliminate the noise is to soundproof the room.

Effort spent in soundproofing is doubly valuable. It not only helps stop undesirable noises from entering the room, it also stops the transmission of sounds produced in the studio to the outside environment. Therefore, with a little soundproofing, it will be possible to work in peace without being interrupted by irate people demanding that the volume be turned down. So read this chapter carefully!

In soundproofing, the general maxim of "You get what you pay for" applies. Sound and vibration are quite devious phenomena and the solutions to controlling them are often intricate and sometimes costly. Nevertheless, with a little care and creativity, it is still possible to do an effective soundproofing job on a small budget.

In this chapter, soundproofing methods have been categorized as follows:
- *for small budgets*
- *for larger budgets*

for small budgets

room openings

Air has a sound transmission coefficient of 1.0 (the highest possible value). Even a tiny air space such as one square inch can let in as much sound as the entire wall adjacent to it. Therefore, large openings in the room, such as doorless entrances or overhead archways, are intolerable. Unnecessary openings should be blocked off with a tight-fitting partition of plywood, gypsum board or similar material. (There are many terms used to refer to drywall construction material: e.g., plasterboard, wallboard, gypsum board, etc. Most of these are acoustically similar.)

In order not to degrade the transmission loss of the surrounding wall, the partition should be contructed from *at least* as many soundproofing layers as the wall in which it is placed. For example, if an opening in a double-layer gypsum wall were to be blocked off, the new partition should be constructed from at least two layers of gypsum—the same construction as the original wall.

Sometimes it is difficult to identify the exact construction of an existing wall. In this case, the new partition should be constructed from a single row of studs with a double layer of gypsum applied to *each* side. The interior cavity of the studs should be lined with fiberglas insulation. Lastly, the joints between the old wall and the new partition should be caulked to prevent the formation of any air leaks.

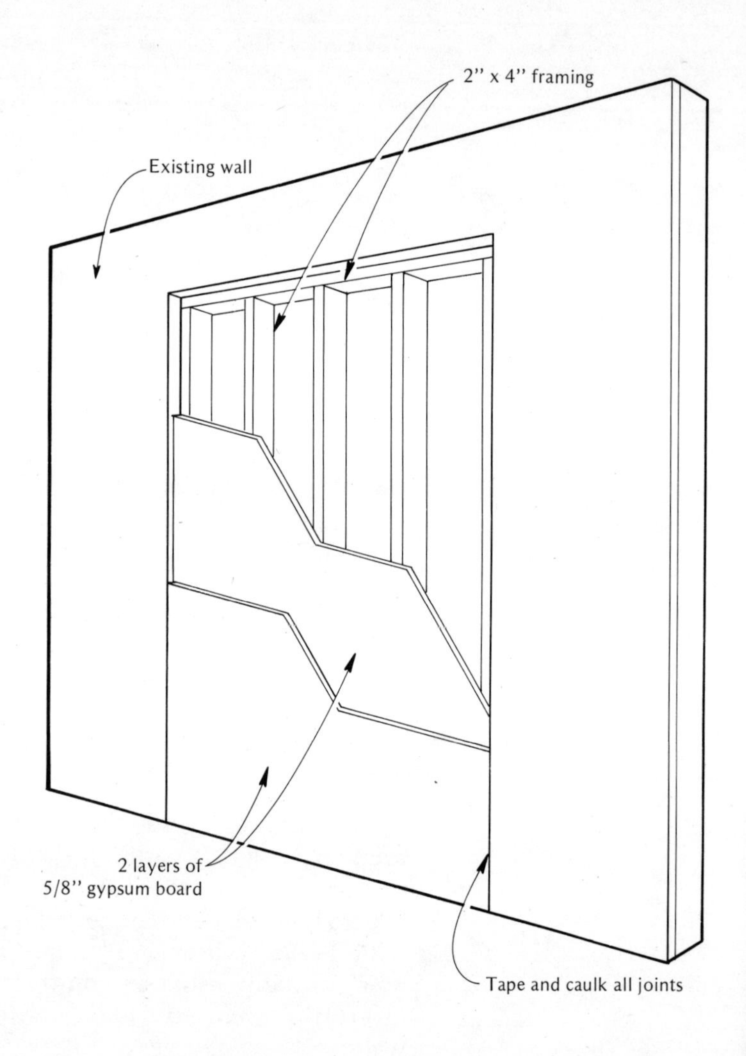

Figure 3-1 Sealing room openings

doors

Other than permanent openings, doors are the most common culprits for sound leakage in rooms. The average interior door found in most homes is a *hollow-core* door that has been *undercut* (shortened slightly) for ease of opening. This type of door is typically constructed of two thin wooden-veneer face panels connected to a frame. Such lightweight, fragile construction offers extremely little resistance to sound.

Figure 3-2 Sound leakage through a typical unsealed door

The average transmission loss (TL) of an unsealed hollow-core door is only 17 dB. (The transmission loss of a typical single-layer gypsum wall would be 35 dB). At low frequencies, such as 50 Hz, the actual TL of the hollow-core door may deteriorate to only 4 or 5 dB.

Sealing the air space around the door by applying weatherstripping or rubber gasketing will help by 2 or 3 dB. However, this is not an appreciable difference.

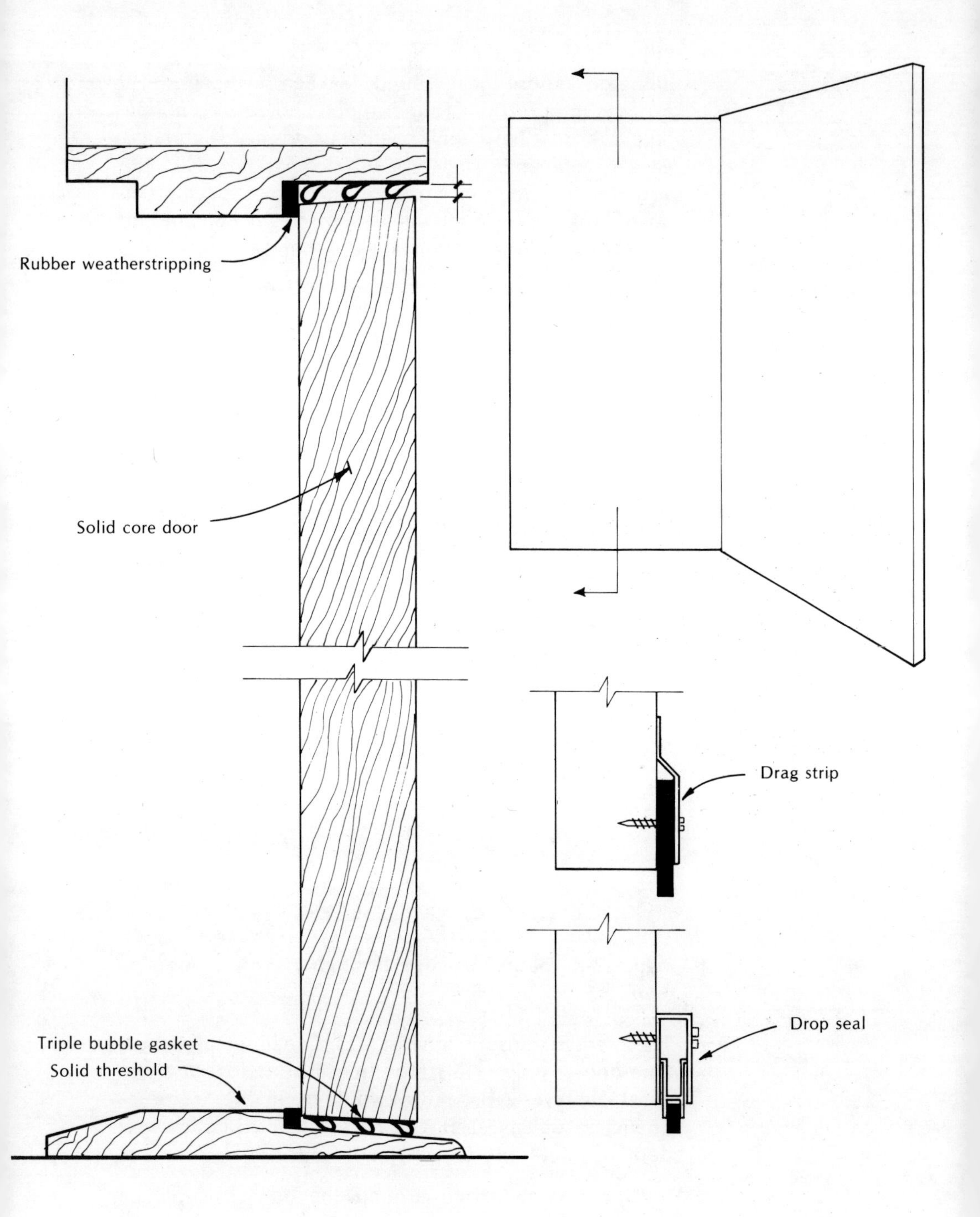

Figure 3-3 Soundproofing Details: Doors

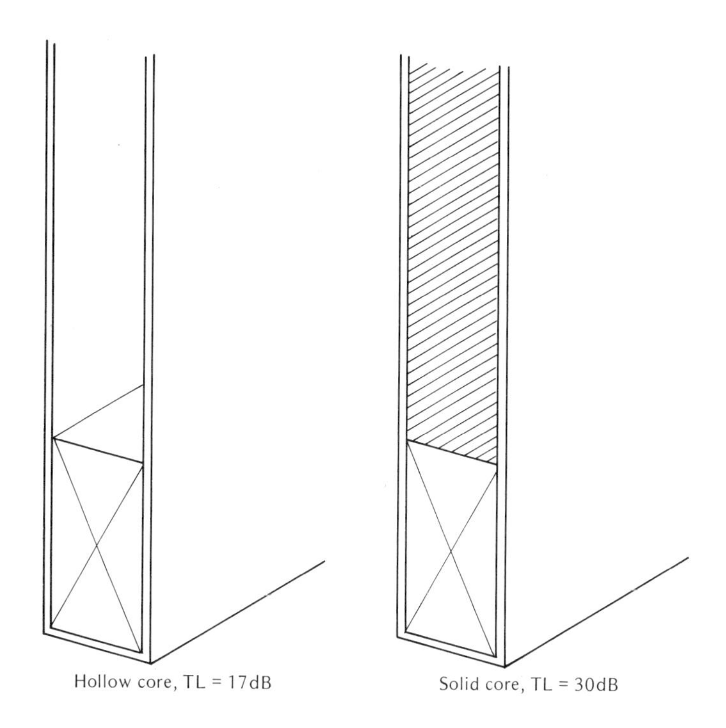

Figure 3-4 Hollow vs. solid core doors

To obtain a distinct improvement, the hollow-core door should be replaced with a more massive *solid-core* door. By applying rubber or felt weatherstripping around the door frame and installing a floor threshold to insure an airtight fit, the average TL of the solid-core door can be brought up to about 30 dB.

Installing a second sealed solid-core door on the other side of the partition will offer the best improvement, bringing the TL up to about 40 dB.

windows

Most windows found in residential and commercial construction are poor sound isolators. The average transmission loss of a typically installed 1/4-inch glass window would be less than 20 dB.

In most cases, this can be raised to approximately 25 dB by sealing the window with a rubber caulking compound, thus preventing air leaks. However, the performance at low frequencies is still lacking.

Figure 3-5 Sound leakage through a typical unsealed window

The least expensive solution is to *block off* the window, in the same manner that other openings in the studio were blocked off. In this way, the TL can be improved to match that of the surrounding wall (typically 30-35 dB).

However, losing the natural light or view from a window is not always desirable. In this case, there are two more intricate solutions. The first would be to build a *hinged* soundproof cover for the window, allowing it to be closed during actual recording. The cover should be rubber gasketed and supplied with a locking mechanism to insure an airtight fit.

A second solution, which makes it possible to keep the natural light during recording, would be to install another pane of glass in the window frame. This pane should be mounted in a sealed enclosure, and separated from the first pane by as large an air space as possible.

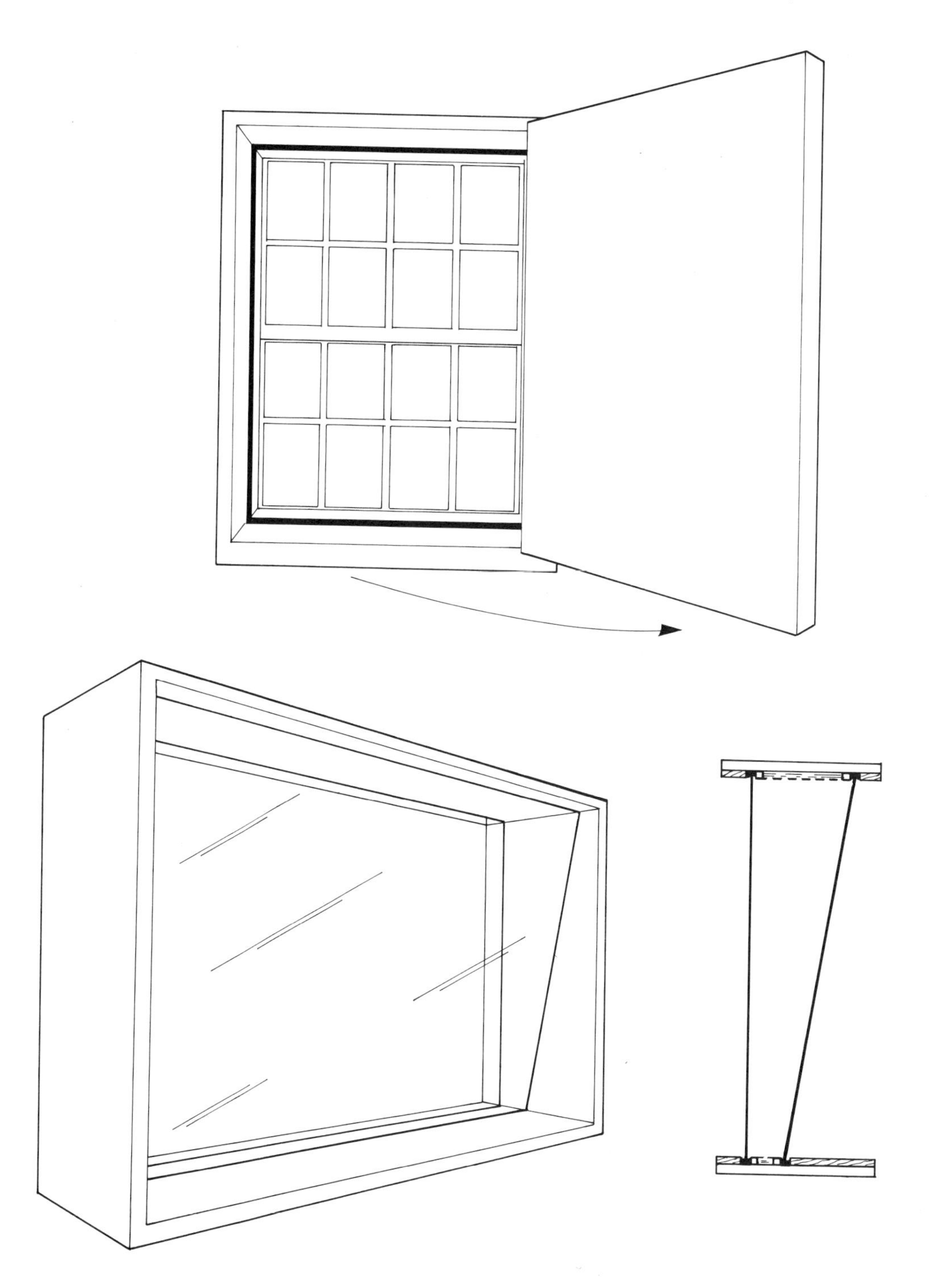

Figure 3-6 Soundproofing Details: Windows

Tilting of the second pane at an angle of 10 degrees, using thicker glass (3/8-inch), and lining the interior window jamb with a sound-absorbing material (such as felt or fiberglas) will increase the performance of the construction appreciably.

A two-pane assembly, fully lined and sealed with a 4-inch air space would have an average performance of 30-35 dB.

Storm windows, often used in northern climates can be utilized for this purpose, providing they are well sealed.

walls

The TL of many walls is severely reduced by poor construction techniques. Unfortunately, wall finishes can hide these deficiencies from view.

It is not uncommon for walls to have dozens of air leaks at the floor and ceiling joints. Sound leaks above false ceilings or through electrical junction boxes are also commonplace.

The first step to improve the performance of an existing wall is to seal the obvious air leaks with caulking compound. If these leaks are not apparent, it is not wise to destroy a wall trying to find them. A more sane approach is to apply a second layer of wall material to the existing wall (5/8-inch gypsum board, for example), thus increasing the total mass of the wall. This can be done on both sides of the partition. By properly sealing the second layer (taping and caulking the joints), the average TL of the wall can be improved by about 5 dB.

An even more effective but slightly more costly method would be to apply spacing strips to the existing wall. These wooden strips (*furring* strips) would be spaced at 16-inch intervals measured from center to center—i.e., "16 inches on center." The space between the strips should be insulated with fiberglas. Then, the new layer of wall material would be applied directly to the spacing strips.

Attached to one side, this type of construction will offer an average improvement of about 6 dB over the original wall. Applied to both sides, the improvement would be about 10 dB.

The following pages contain diagrams of various wall and floor constructions. Each construction has been rated for its soundproofing ability by utilizing the STC rating system.

The STC System

STC (Sound Transmission Class) is a single number rating system commonly used for comparing the soundproofing characteristics of various constructions. The STC system matches a given wall's tested acoustical performance to the most closely matching Standard Contour, using a detailed set of rules. Standard STC contours slope upward in the low frequencies. It should be noted that the STC system tends to overrate lightweight walls which have good attenuation at high and mid frequencies, while underrating heavy walls with good attenuation at low frequencies.

In studio construction, low frequency attenuation is of prime importance. Therefore, the test results of each wall shown have been categorized into 5 dB ranges. These ranges give a truer picture of low frequency performance and allow for deviations between lab tests and actual field performance due to imperfect seals and flanking as well as small variations in construction technique.

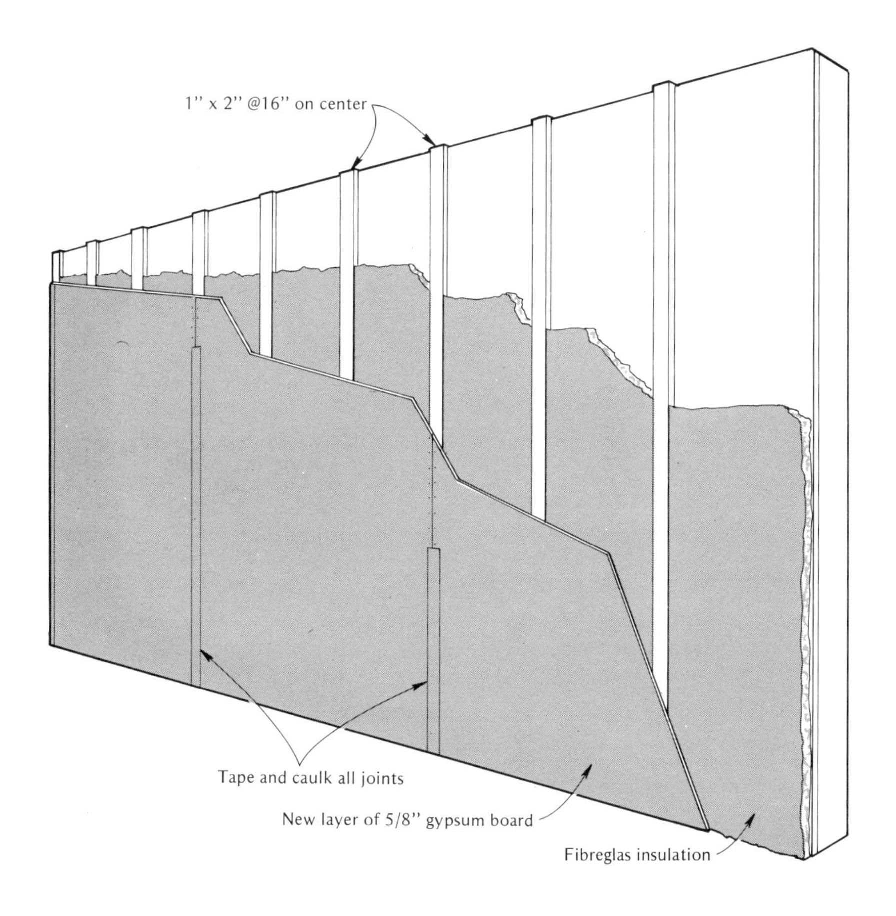

Figure 3-7(a) Adding a new layer to the wall

STC 30-35

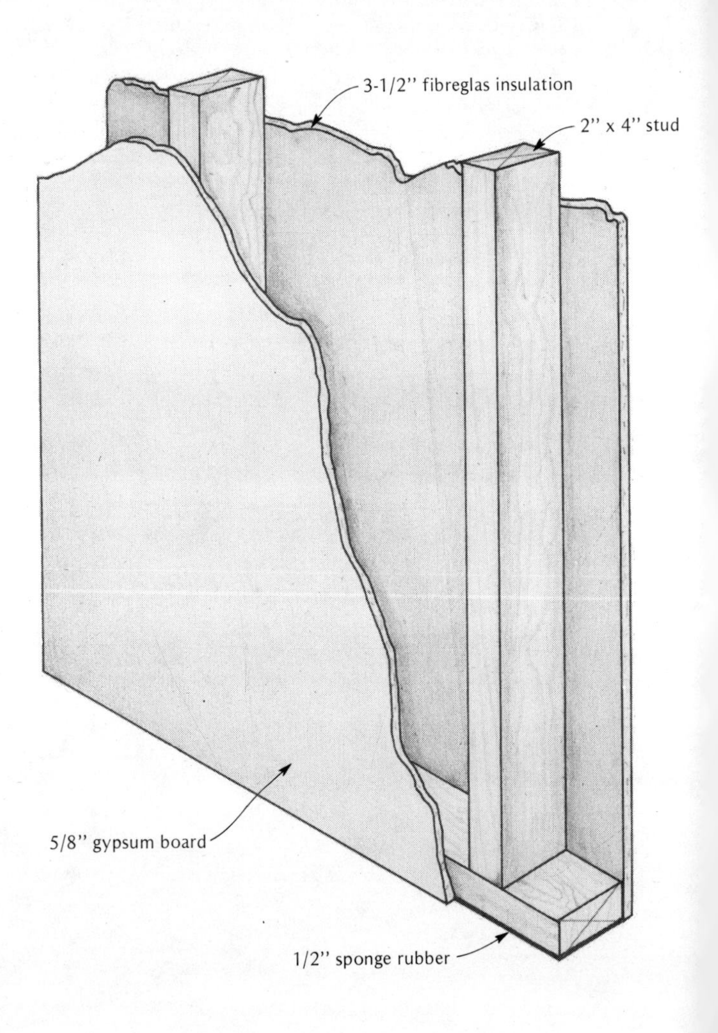

Figure 3-7(b) Standard drywall partition

STC 35-40

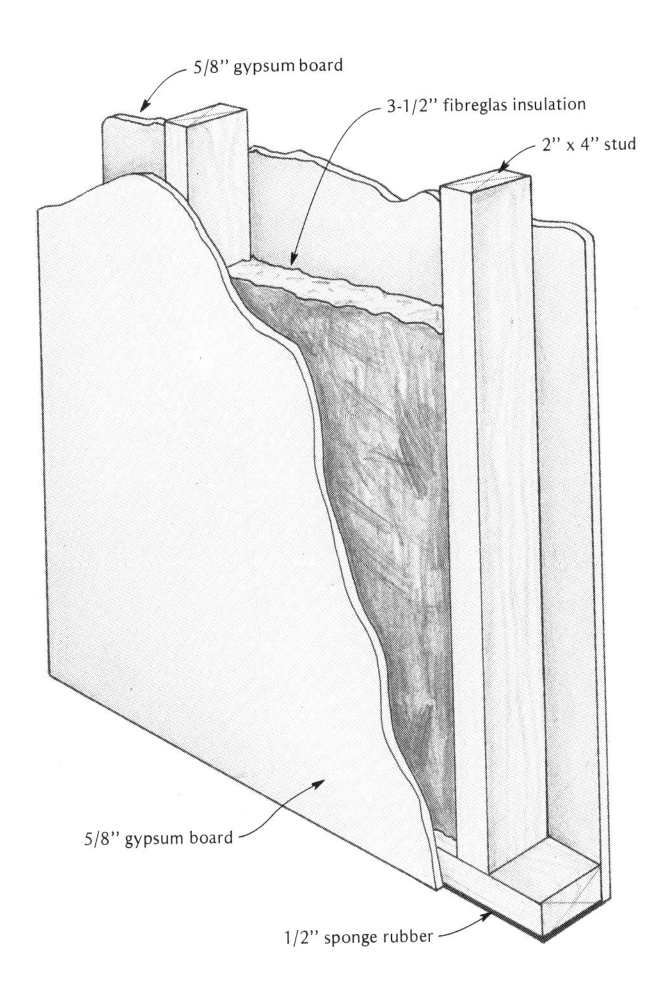

Figure 3-7(c) Standard drywall partition with insulation

STC 40-45

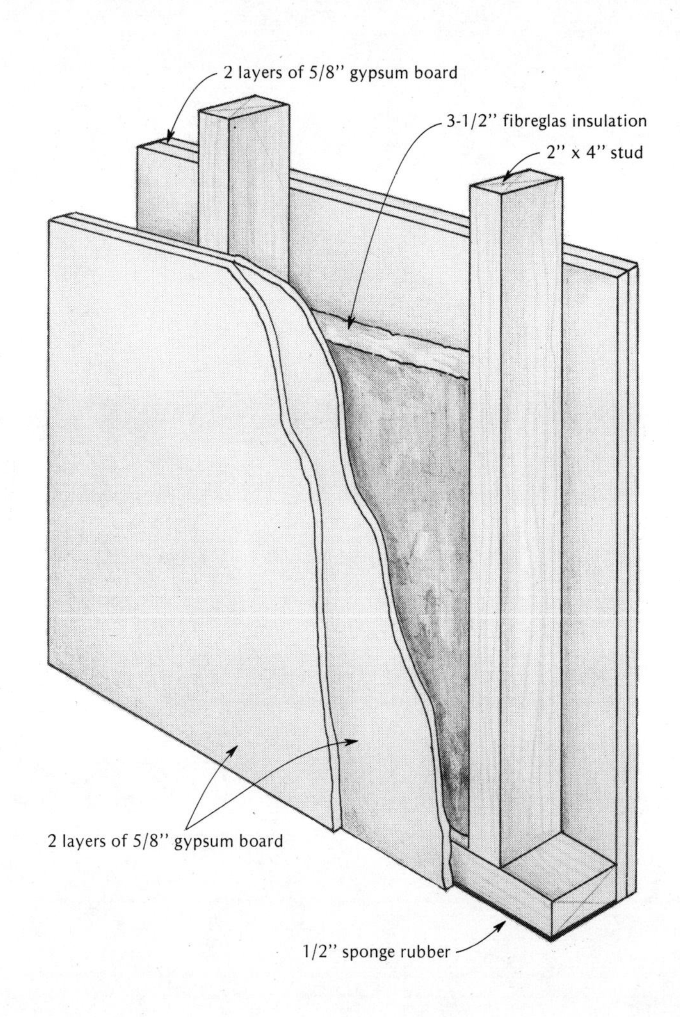

Figure 3-7(d) Double layer drywall partition

STC 40-45

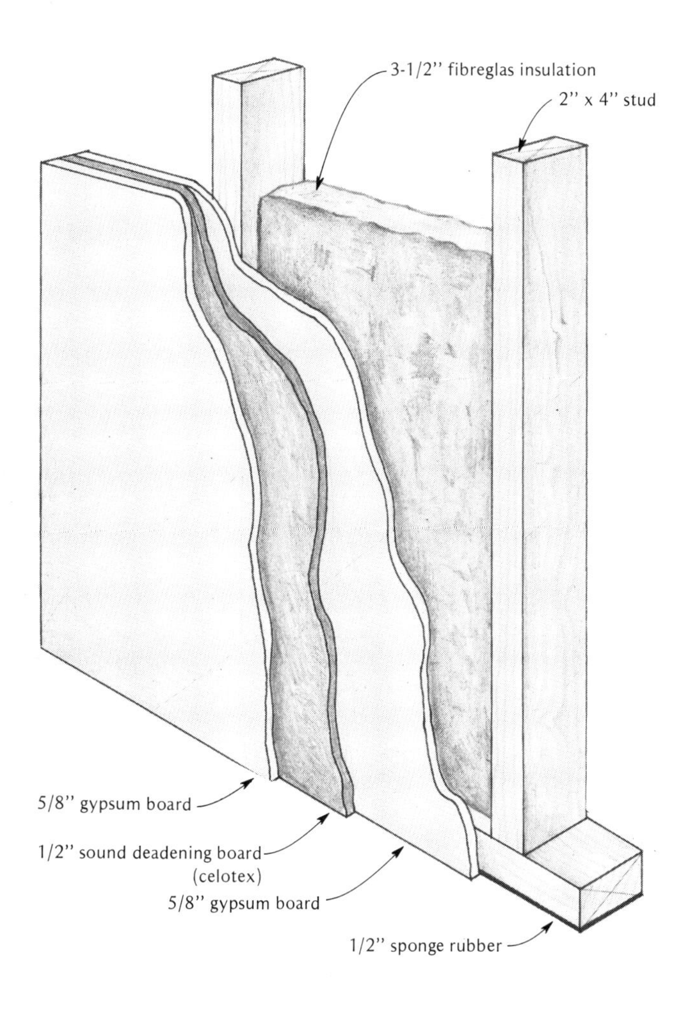

Figure 3-7(e) Resilient sandwich partition (one side only)

STC 45-50

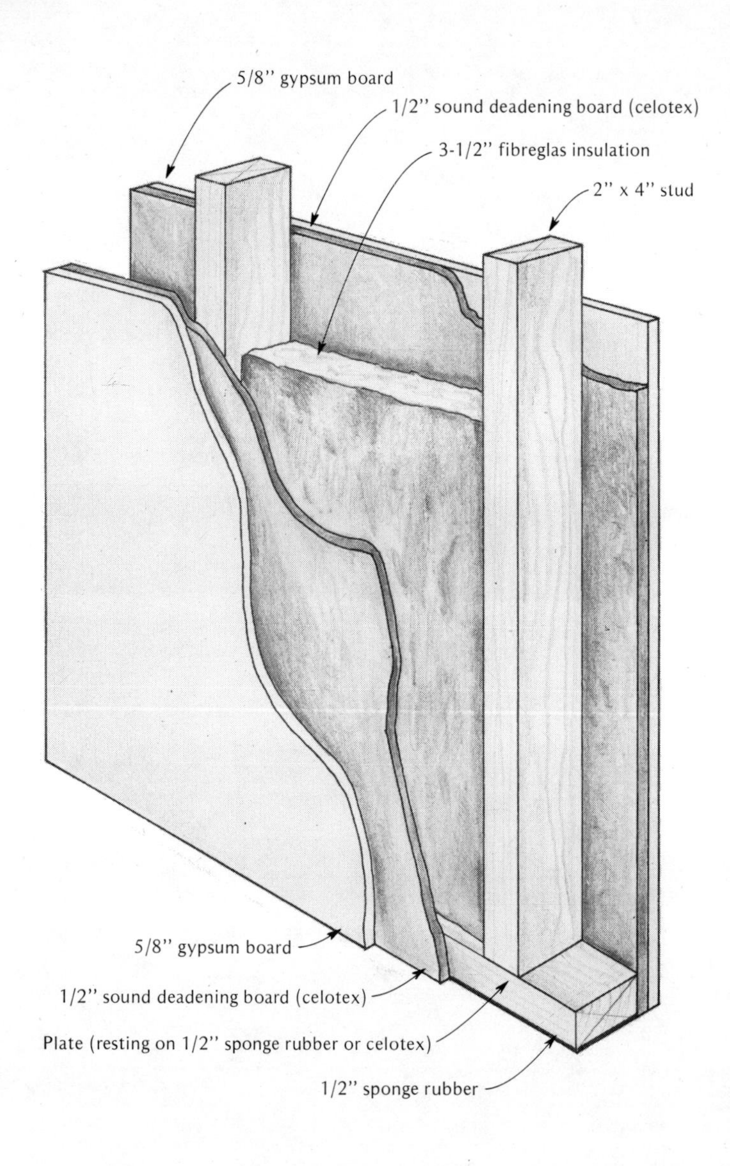

Figure 3-7(f) Double-sided resilient sandwich partition

STC 50-55

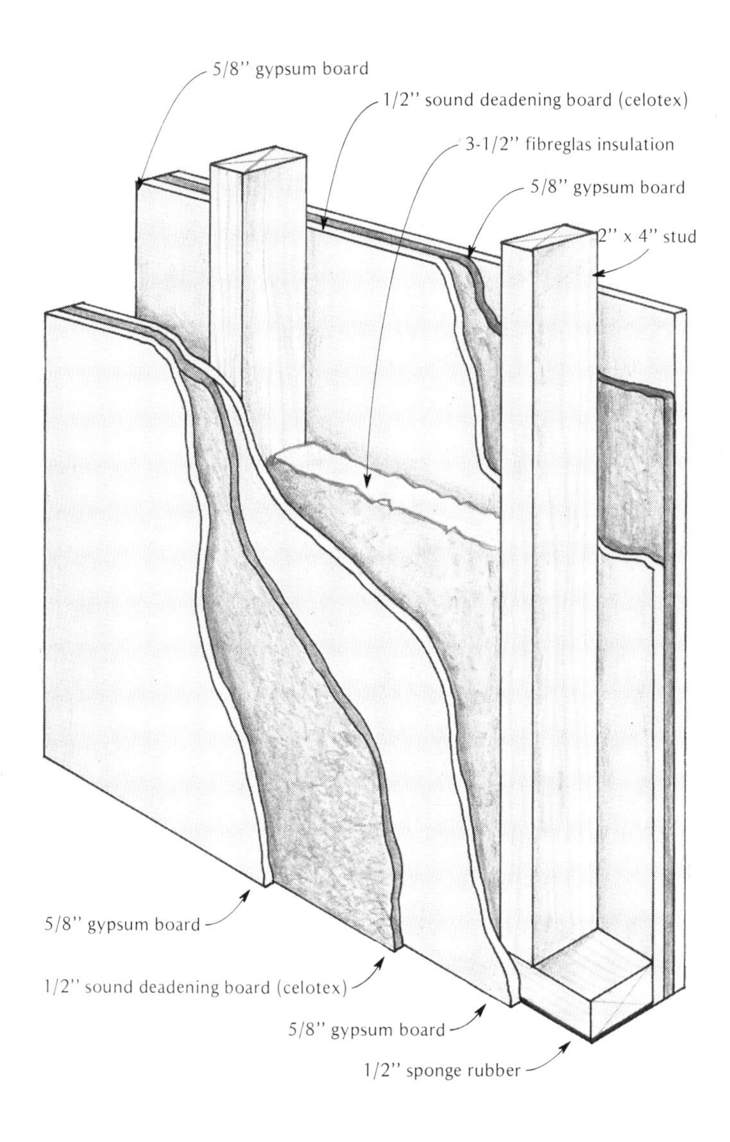

Figure 3-7(g) Double-sided double layer resilient sandwich partition

STC 55-60

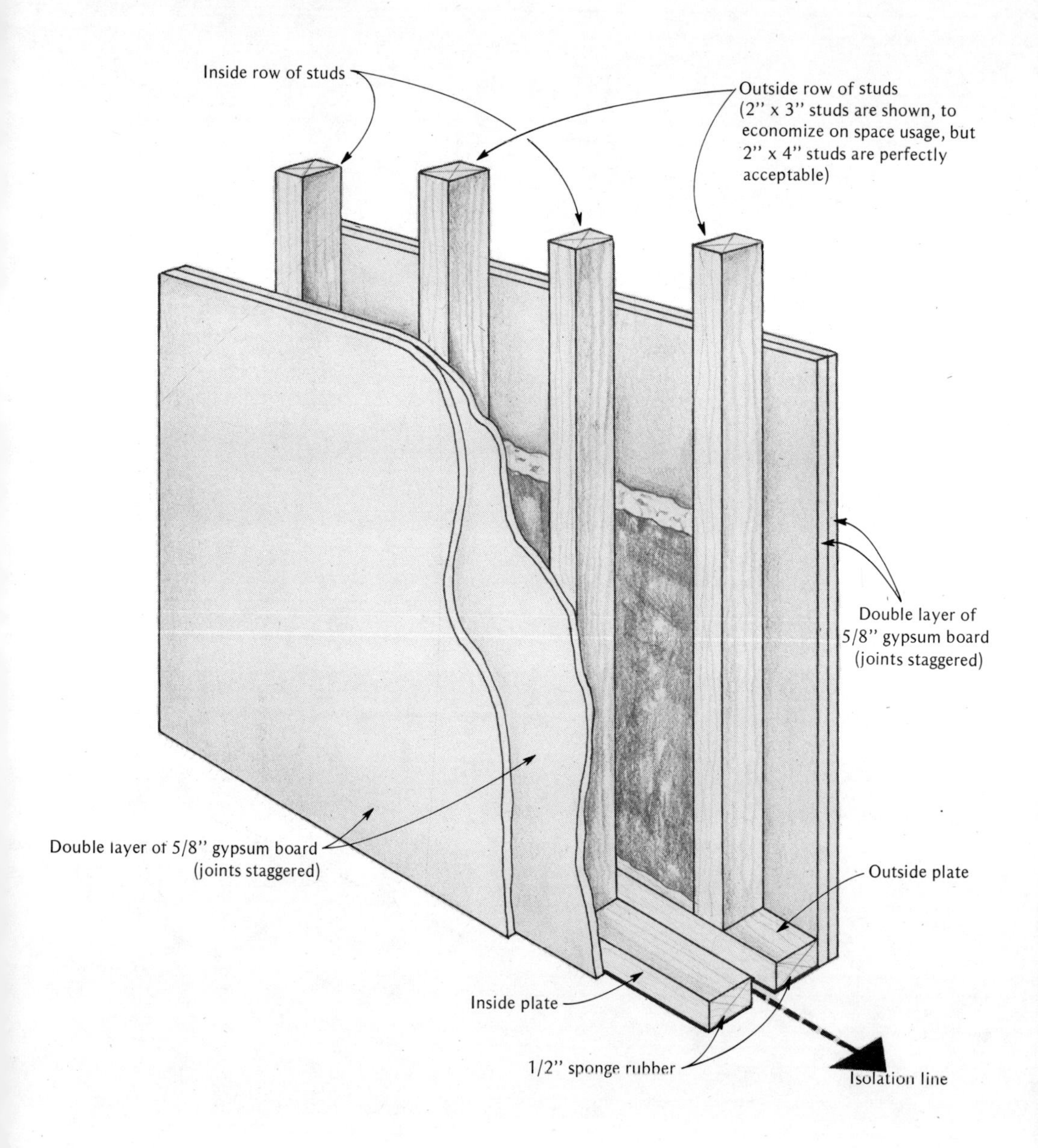

Figure 3-7(h) Double-wall construction

STC 50-55

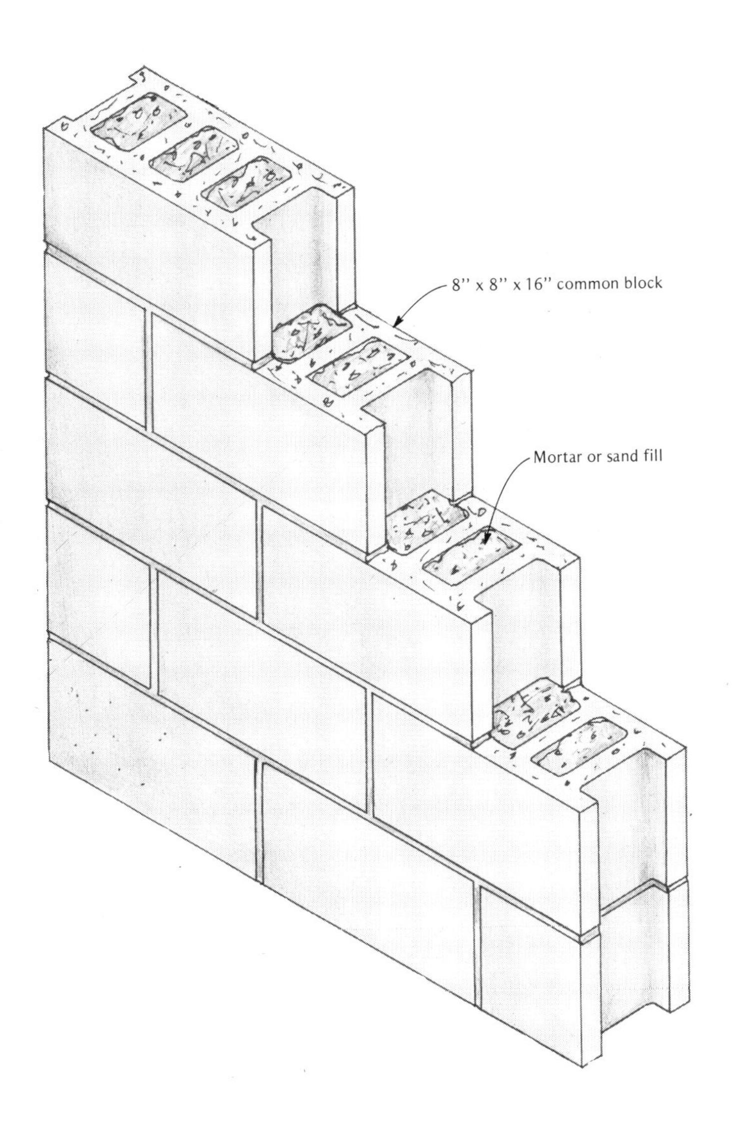

Figure 3-7(i) Concrete block wall

STC 60-65

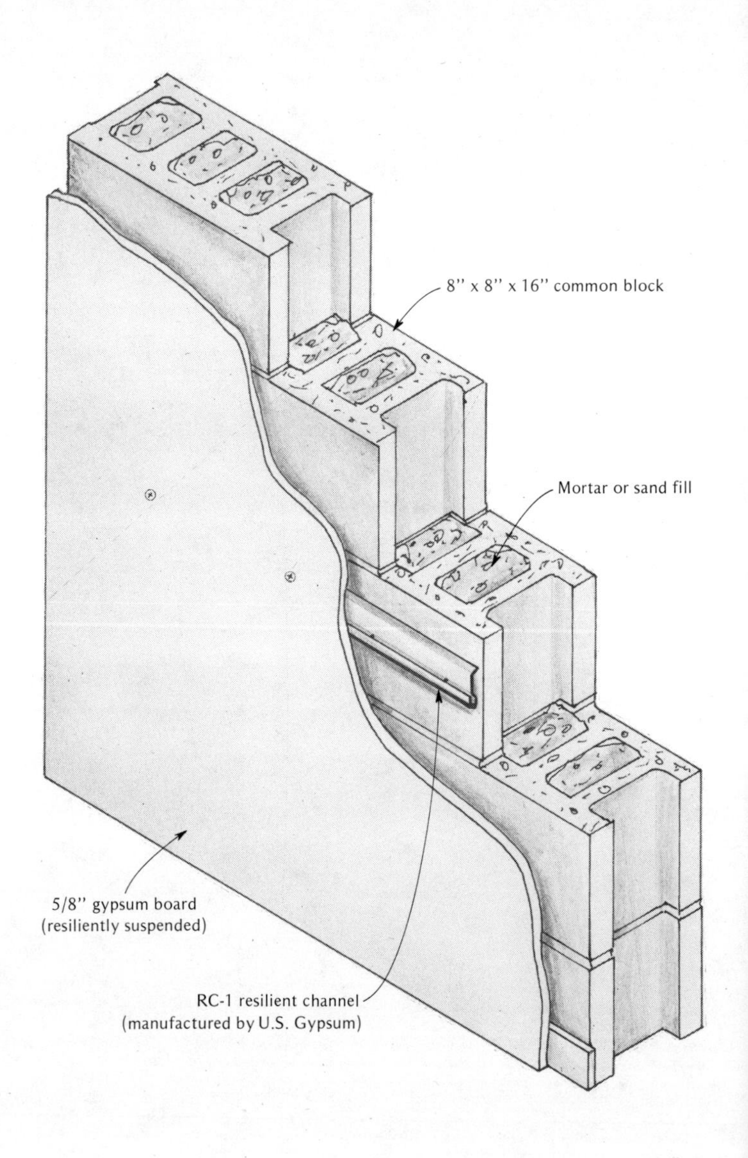

Figure 3-7(j) Concrete block wall with resilient surface skin

STC 65-70

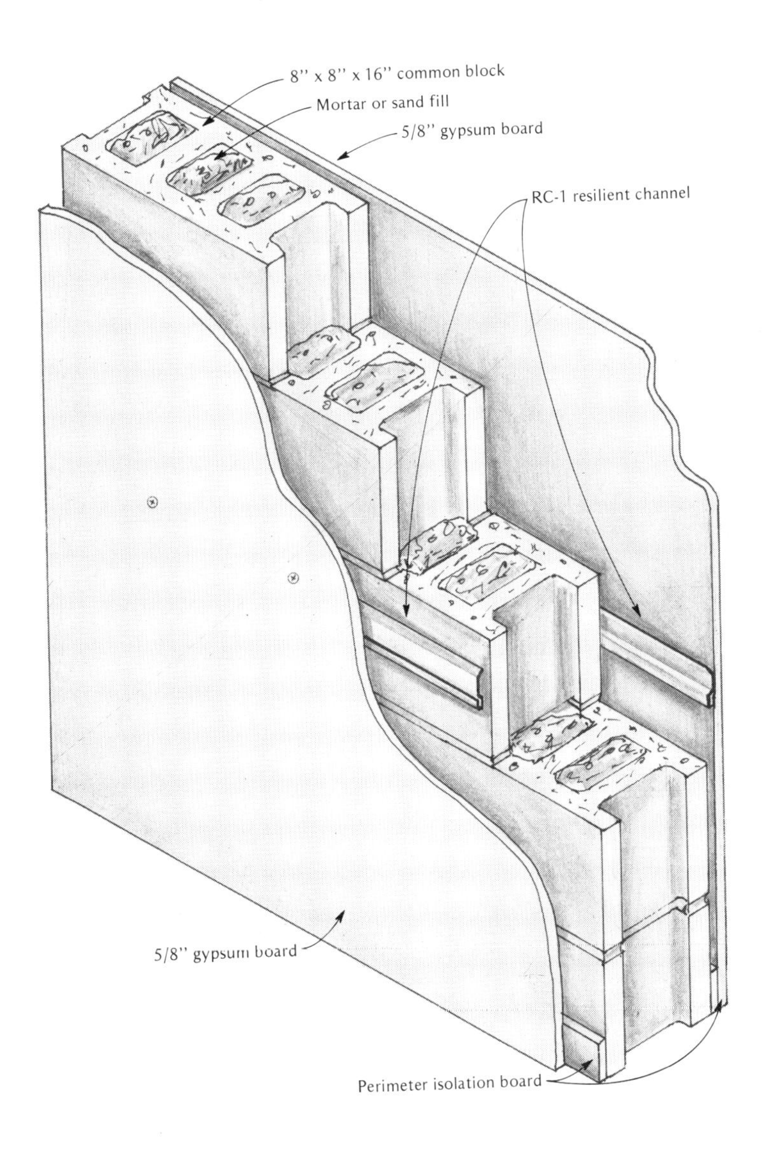

Figure 3-7(k) Concrete block wall with resilient surface skin both sides

ceilings

Since ceilings are often used to hide ductwork, wiring and light fixtures, they provide ample opportunity for the penetration of sound.

The first step towards improving the TL of an existing ceiling is to seal all air spaces with caulking compound. This can usually provide an improvement of several dB. If the ceiling is unfinished (i.e., the decking of the roof or floor above is visible), the underside should be closed off with a double layer of drywall or similar material and the air space above lined with a layer of fiberglas insulation. These measures will provide an improvement of 10 to 20 dB.

If the existing ceiling has already been finished (i.e., plastered or drywalled), its TL can be further improved by the methods described for walls in the previous section. These would include the sealing of existing leaks and the application of additional ceiling layers, either directly or via the use of spacing strips and an insulated air space. Such techniques would increase the TL of the ceiling by an additional 5 to 10 dB.

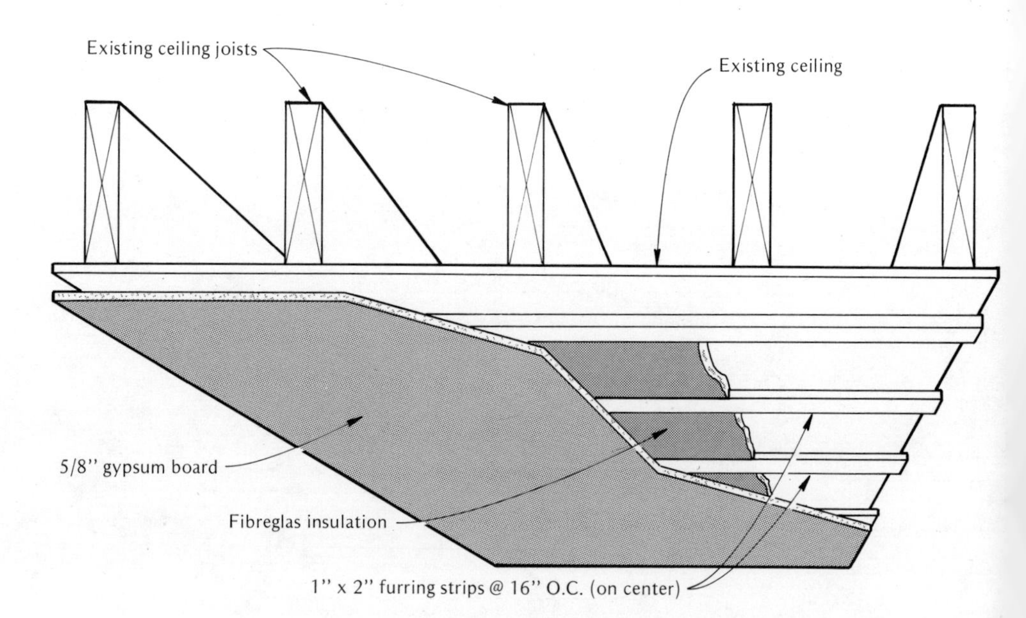

Figure 3-8 Adding a new layer to the ceiling

These soundproofing methods will chiefly affect *airborne noise* such as noise from airplanes and close traffic. However, many ceilings are subject to another type of noise called *impact noise*. Impact noise refers to vibrational disturbances carried directly through the structure of the building itself and includes such sounds as foot thumps and furniture scrapes, etc. Because of the innate ability of most structural materials such as wood and steel to carry vibrational impulses, impact noise is much more difficult to control than airborne noise. As a result, such techniques as sealing air spaces and adding new soundproofing layers do not always reduce impact noise to the desired level.

The least expensive solution to controlling impact noise from above is to apply a carpet and thick underpadding to the floor over the studio. This helps decrease the level of impact sounds at the source, before these sounds actually enter the structure.

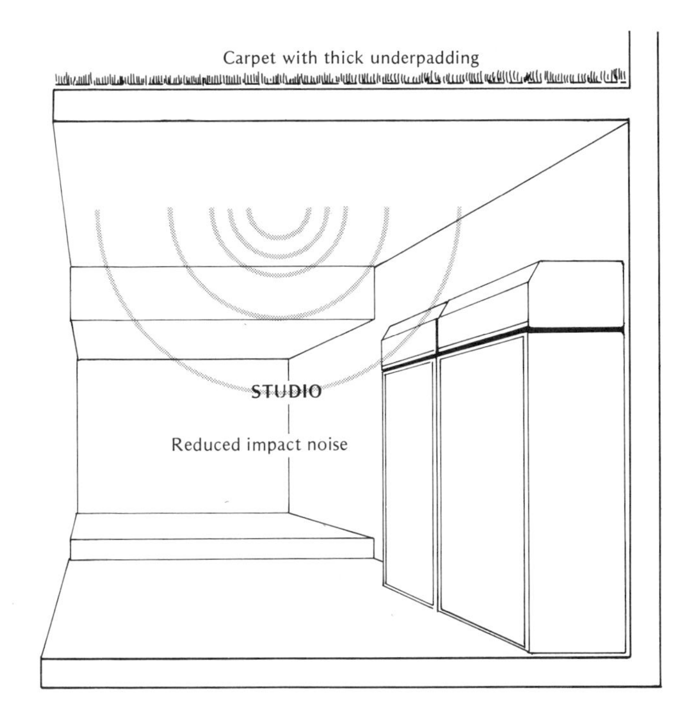

Figure 3-9 Reducing impact noise from above

Although carpeting the floor above will sometimes prove effective enough, the ultimate solution to controlling overhead impact noise involves building a new ceiling suspended on vibrational isolators. This is a rather intricate solution and it is discussed in a later section (*for larger budgets*).

It should be noted that the so-called "acoustical ceiling" often found in homes and office buildings does little or

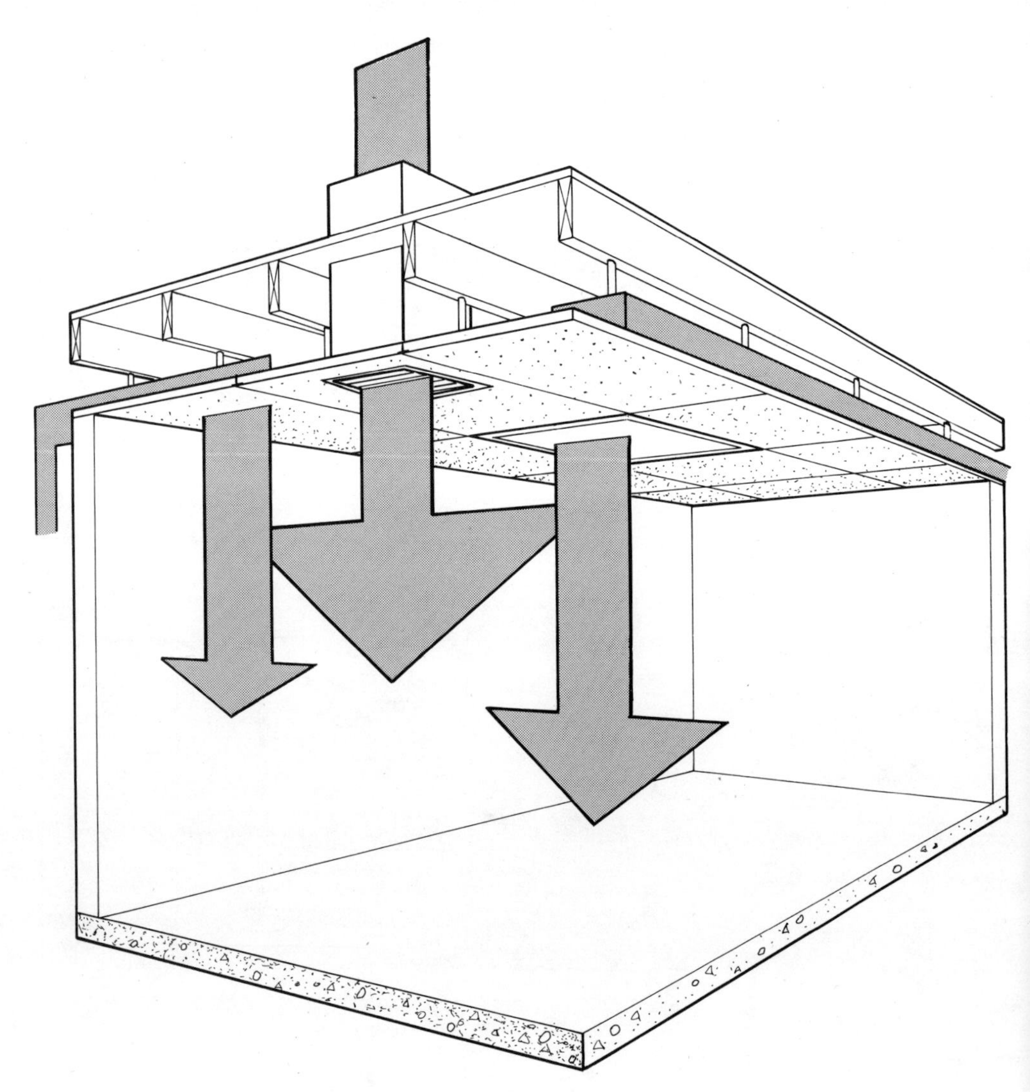

Figure 3-10 Sound leakage through a typical acoustical tile ceiling

nothing to stop airborne or impact noise. This type of ceiling is normally constructed from a metal framework and a grid of lightweight "acoustical tiles." As a result of its fragile construction containing numerous penetrations and air spaces, "acoustical ceilings" offer little more than cosmetic value for serious soundproofing. The main purpose of such a ceiling is to provide additional absorption in the room in which it is hung, thus shortening the reverberation time and giving the room a *dead* sound.

floors

The soundproofing of studio floors is a difficult and often costly task owing to the high impact and noise levels to which a studio floor is normally subjected (bass drum kicking, foot stomping, etc.). The most effective solutions involve building a new floor and, as a result, are rather intricate; they will be dealt with in the next section (*for larger budgets*).

The application of a carpet, a solution often tried, does little or nothing to affect the transmission of airborne sound. Airborne sounds from pianos, electric instruments, monitor speakers, and even voices travel right through a carpet because of its lightweight and porous composition. Furthermore, a carpet does little to stop the large impact noise commonly experienced in studios. For these reasons, carpeting a studio floor has almost no value for soundproofing.

The principle effect of the carpet is to stop sound reflections *in* the room, thereby shortening the reverberation time at mid and high frequencies, giving the room a *dead* sound.

Sometimes, carpeting the room *above* a studio helps reduce the transmission of lesser impact noises from above (such as footsteps) into the recording area. This is recommended if such noise is a problem. It is much easier to stop impact noises *before* they enter the structure than afterwards.

The following pages contain diagrams of various floor constructions. Each construction has been rated for its soundproofing ability.

Utilizing the same rating system previously discussed for wall construction, the STC System, each floor has been graded in 5 dB ranges. This range better approximates a given floor's performance in the lower frequencies (below 250 Hz), since this range is critical to the soundproofing of the studio and control room.

Impact Noise Rating

STC ratings refer only to airborne sound transmission. It is important to note that in addition to airborne noise, floors are also subject to impact or structureborne noise from such occurrences as footfalls, mechanical equipment vibrations, nearby road vibrations and similar physical disturbances.

For that reason, the ***Impact Noise Rating (INR)*** *of each floor is also given in the accompanying diagrams. The INR is a rating system similar to the STC System. Both compare the tested impact isolation performance of a given floor to a standard group of contours.*

Negative ratings, such as −30 to −25, imply relatively poor impact isolation. Positive ratings, such as +28 to +33, imply relatively good performance. Ratings in the 0 to +5 range imply rather standard performance.

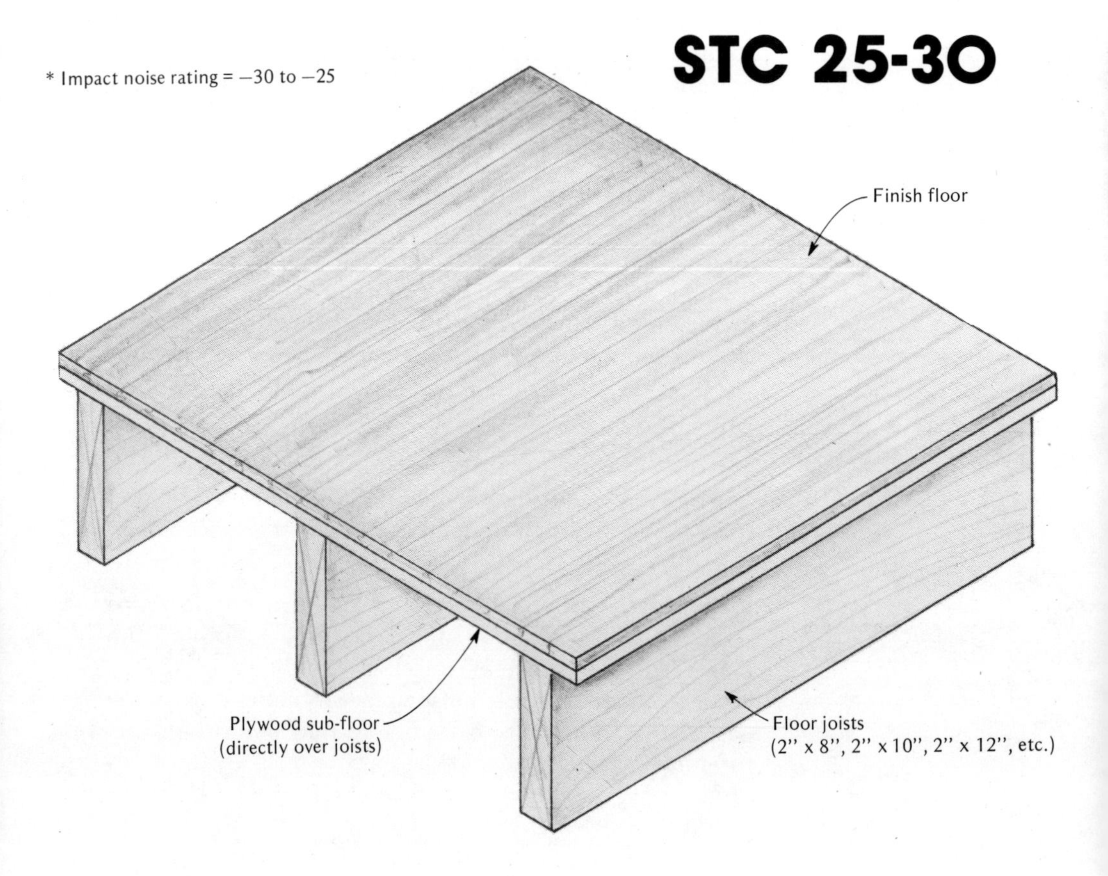

Figure 3-11(a) Common wood joist floor

STC 37-42

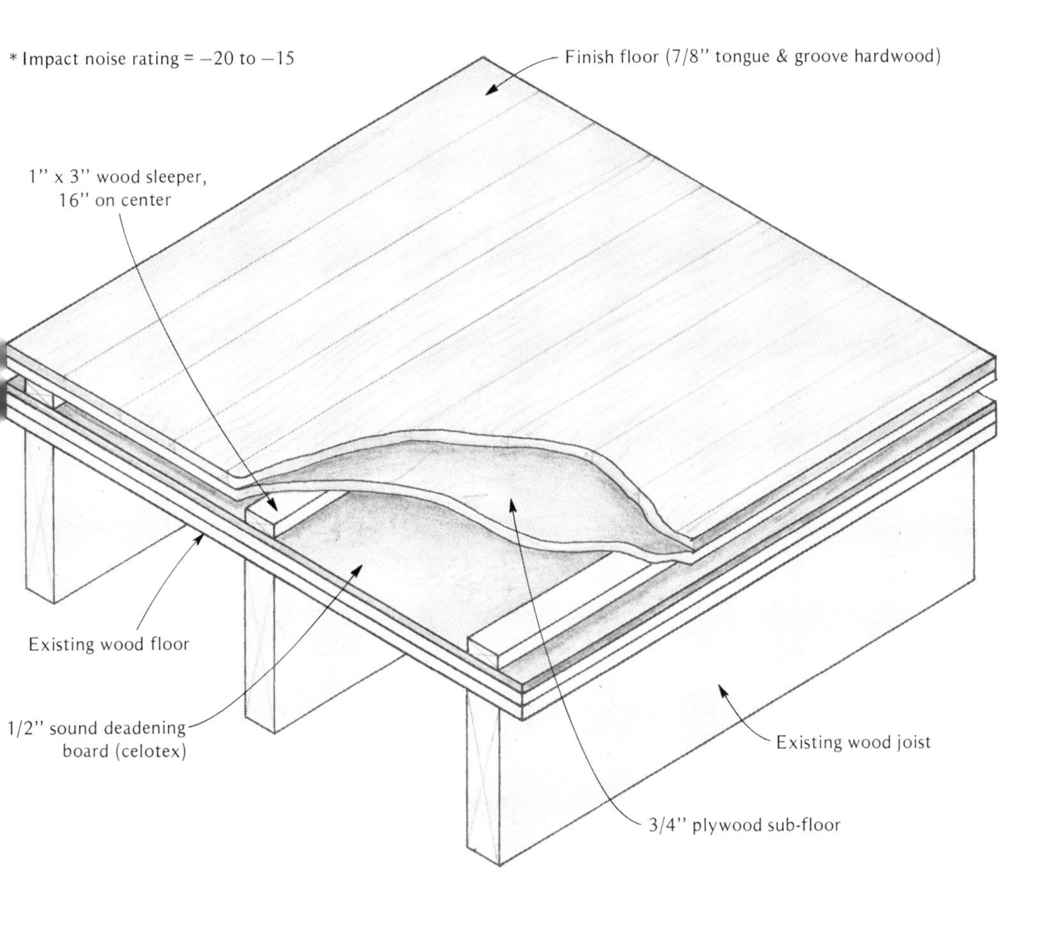

Figure 3-11(b) A simple wood floating floor

STC 57-62

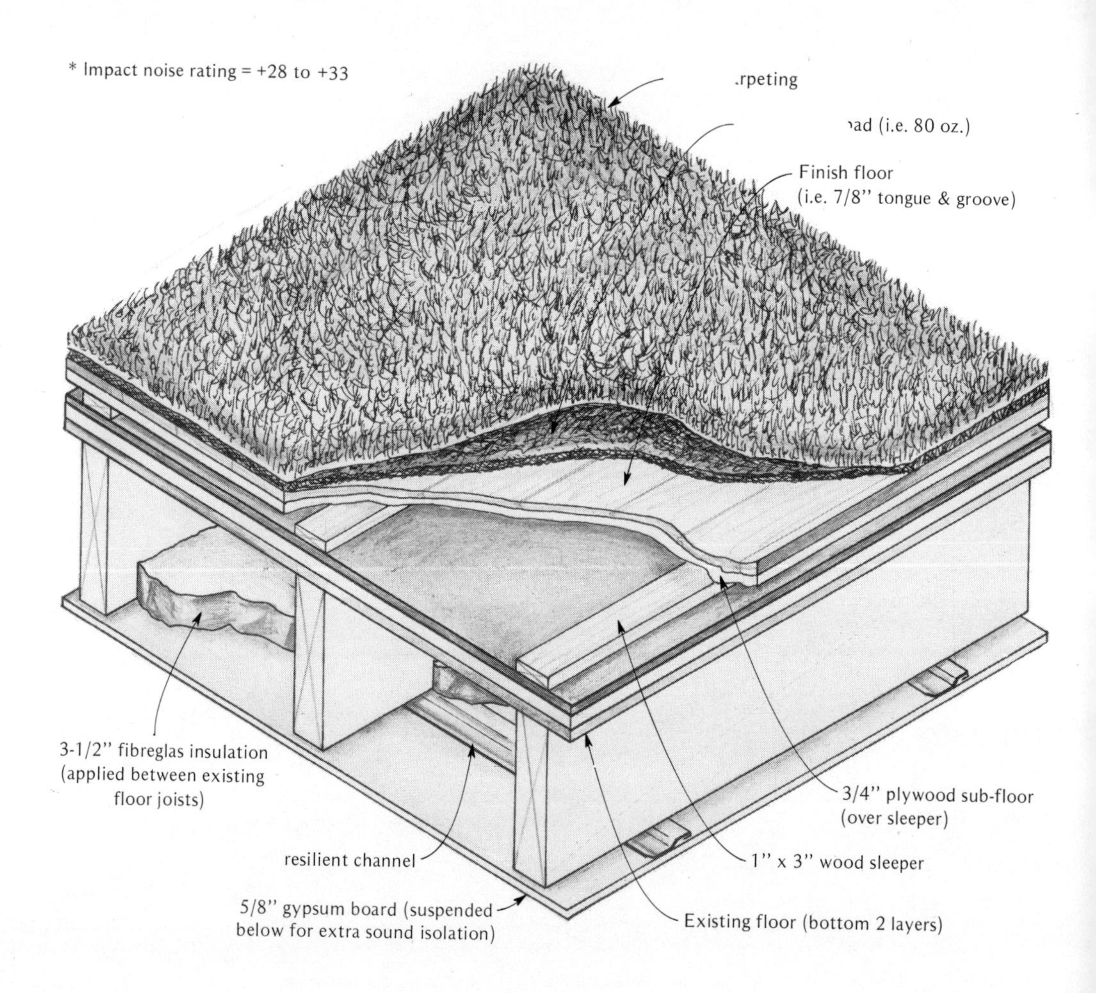

Figure 3-11(c) A higher performance wood floating floor with suspended ceiling attached

STC 45-50

* Impact noise rating = −20 to −15

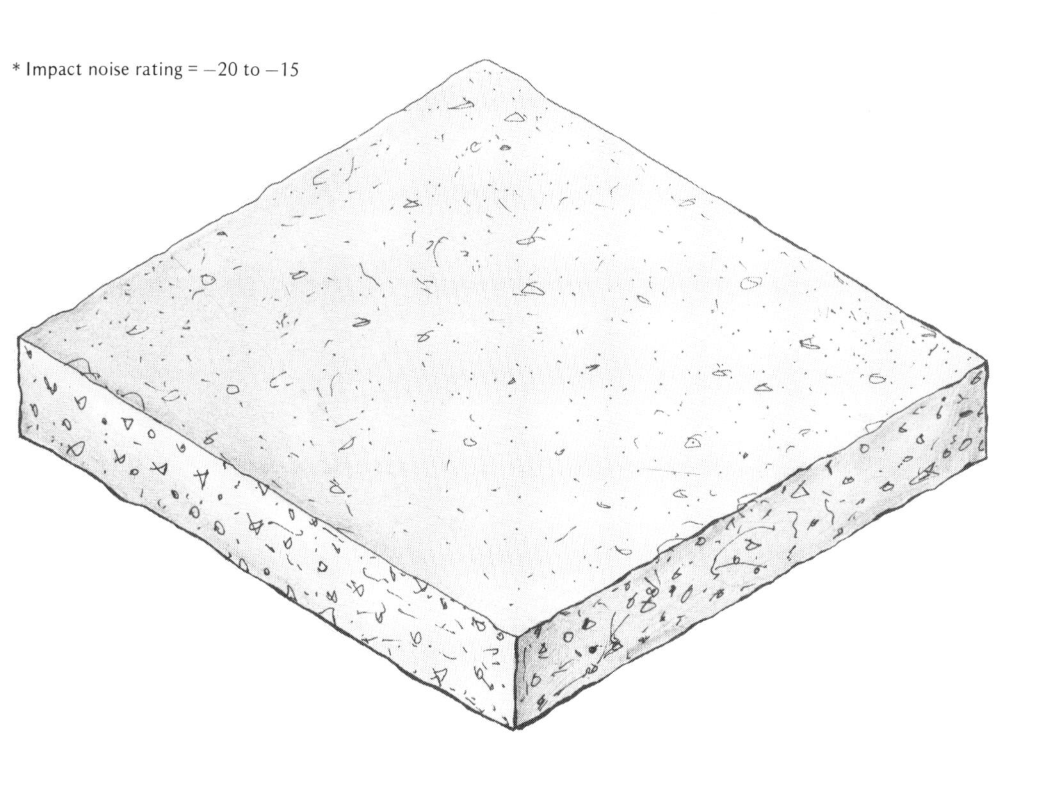

Figure 3-11(d) Standard concrete slab

STC 45-50

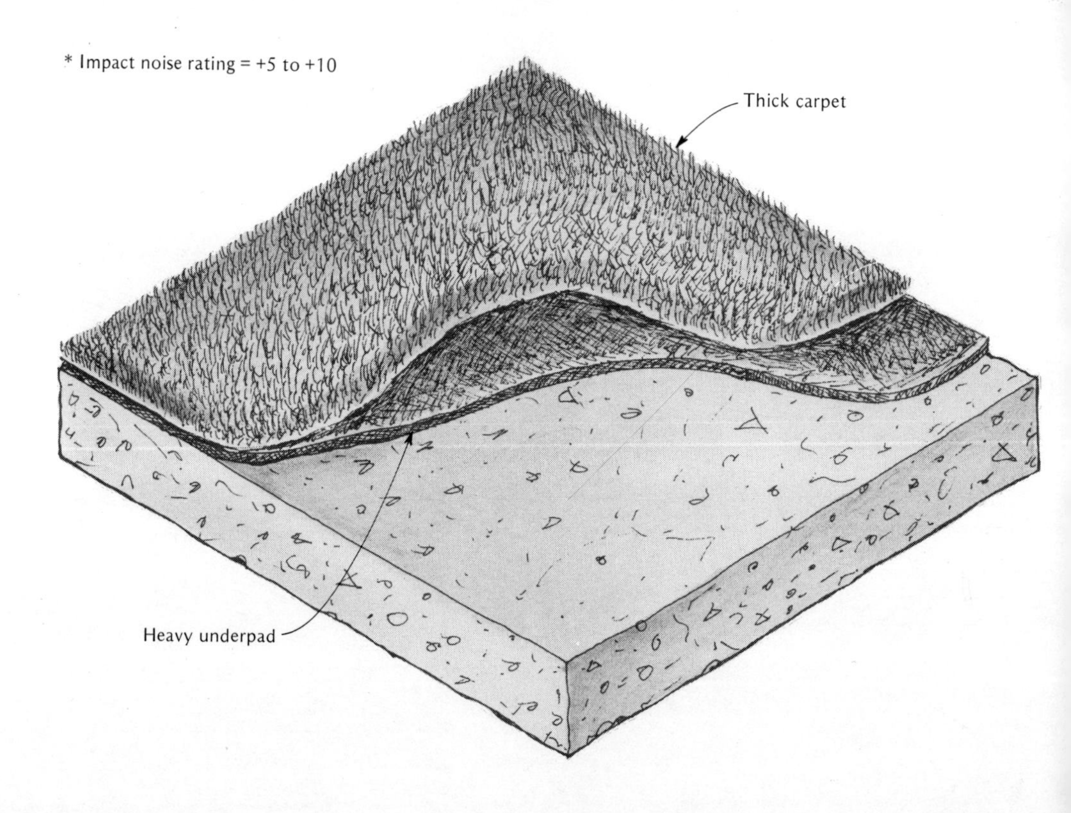

Figure 3-11(e) Standard concrete slab with thick carpet and underpadding

loose objects

Loose objects, such as paintings, wall hangings, sculptures, lamps and small tables, have a tendency to resonate at their natural frequency when loud music is played. These resonances often produce humming and buzzing sounds that can easily be picked up by microphones, thereby ruining a recording.

Loose objects should be firmly mounted or eliminated entirely.

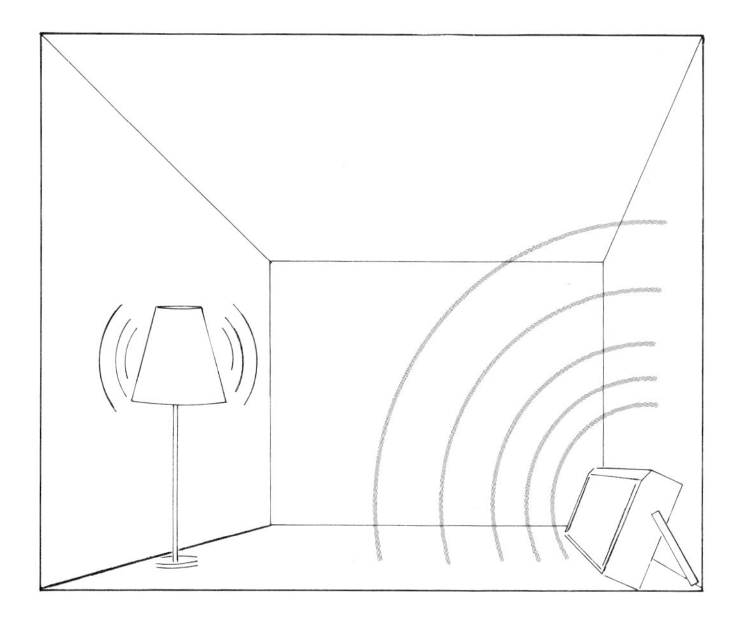

Figure 3-12 Sympathetic resonance

appliances

Air conditioners, heaters and other appliances can produce enough extraneous noise to ruin an otherwise good recording. If the operation of the appliance is not strictly essential, it should be removed or turned off during recording. Appliances that cannot be removed or disconnected should be acoustically isolated. Any appliance with a heavy motor or fan (e.g., air-conditioning unit) can be mounted on springs or a rubber pad to help reduce transfer of vibrations to the immediate structure.

An additional measure is to build a soundproof enclosure, lined with fiberglas, that surrounds the noise-producing equipment.

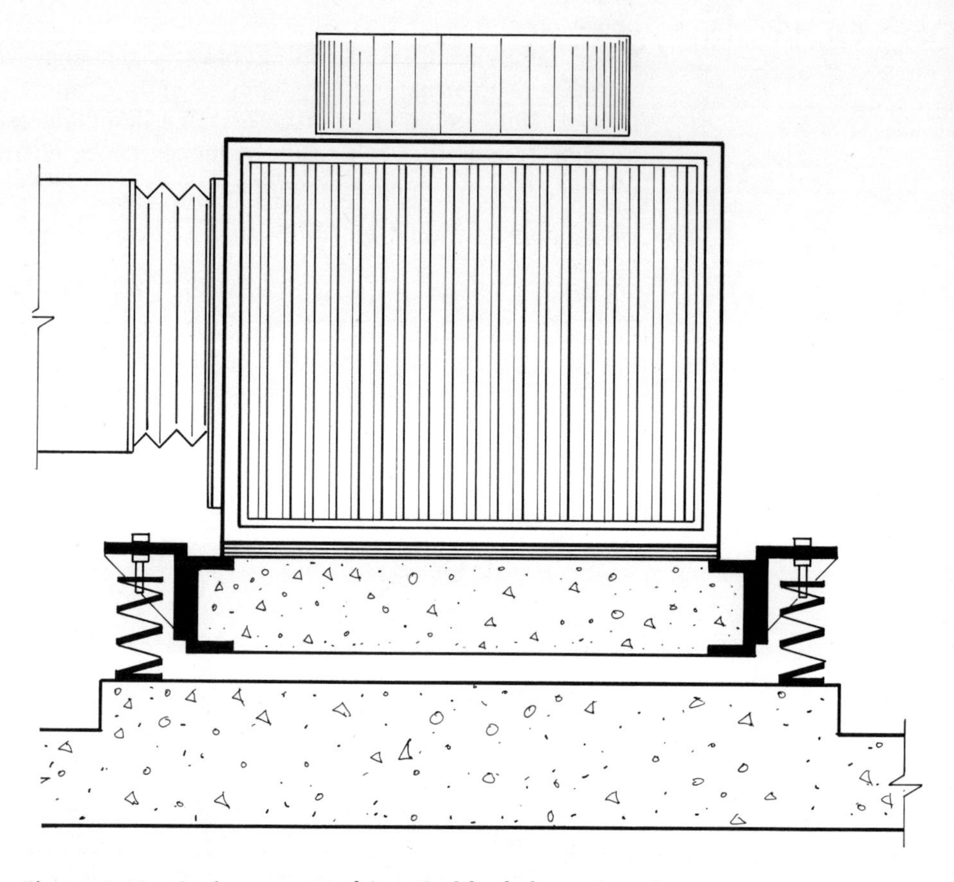

Figure 3-13 Spring mounted inertia block for noisy air conditioning unit

don't add absorption to your room

Adding absorption to your room is *not* the way to soundproof it. Although covering a room with carpet and acoustic tile might reduce the level of outside noise heard in the room by a few decibels (perhaps 3 to 5 dB), it also makes the room seem acoustically dead—a terrible environment for playing music. The acoustics of the room matter too much. Don't cover your room with acoustic tile and carpet in hopes of reducing noise transmission! This is *not* the way to design a good-sounding studio. The proper method is discussed in the next chapter.

for larger budgets

introduction

The soundproofing methods described in the preceding section fall generally into the category of *tightening and fortifying* existing construction (by sealing leaks, insulating air spaces, increasing the mass of existing walls, lining ductwork, etc.). These relatively inexpensive methods will work satisfactorily up to a point, but cannot be expected to provide *complete* soundproofing of the studio from its immediate physical environment.

Sound is a vibratory phenomenon. As long as the studio or the control room is *rigidly* connected to the immediate environment, the transfer of sound waves and vibration to and from that environment is inevitable. As a result, there is a point of diminishing returns beyond which further sealing and insulation of the studio (or control room) will not increase its acoustical isolation. At this point, the most effective way to proceed is to physically separate the studio from its peripheral surroundings.

By separating or *floating* the construction of the studio from its immediate surroundings, we can effectively impede the transfer of sound and vibration to and from the studio. According to the *mass law* of physics, a solid barrier's ability to stop sound transmission increases by approximately 5 dB each time its weight is doubled. However, as much as *three times* the predicted isolation of an equivalently massive structure can be obtained by using the floating principle.

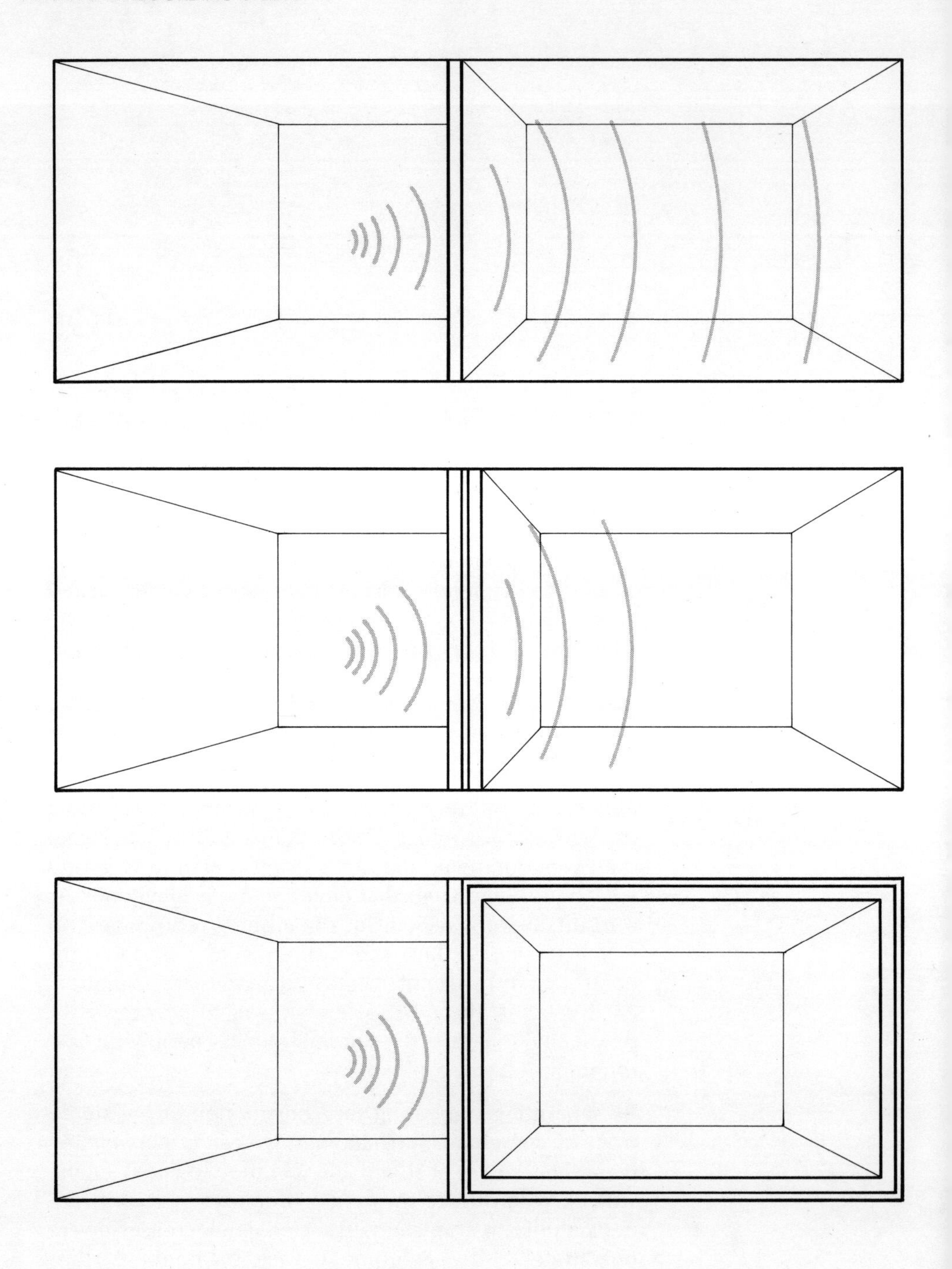

Figure 3-14 The floating studio

Specifically, a *floating studio is one whose walls, ceiling and floor are separated from the exterior architectural shell by an insulating air space.* Since a floating studio is actually an entire new room built within the existing structure, we will examine its construction by considering, in turn, each of its components. These are:

- walls
- floors
- ceiling
- doors
- windows
- ventilation
- wiring

The details are applicable to both the *studio* and the *control room* construction.

the floating studio

walls

Noise transmission to and from adjacent spaces can be reduced considerably by building floating walls separated by an air space from the existing wall. It is important that these new walls be resilient—i.e., not rigidly connected to any part of the existing structure. This prevents vibrations from being transferred from the inner walls to the outer walls and vice versa. This resiliency is accomplished by supporting the walls on *vibration-isolation mounting board* or by hanging *resilient clips*. The wall itself can be made from plywood, plaster board, gypsum drywall material or even masonry brick—if the existing structure can support the load. The walls should completely surround the entire studio area.

The floating walls should not contact the outer walls *at any point*. The new walls should be caulked to eliminate any possible air leaks. The space between the floating walls and the existing structure should be lined with porous absorbent such as fiberglas. These details are illustrated in the accompanying diagram.

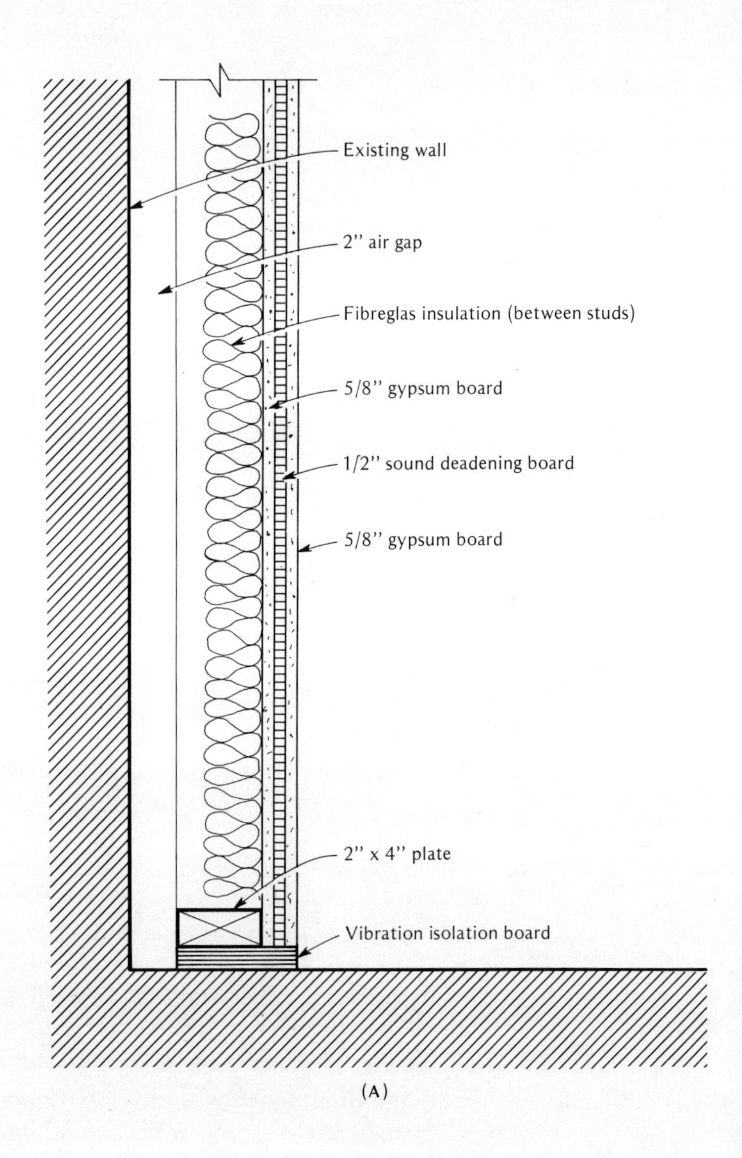

Figure 3-15 A typical floating wall (wood)

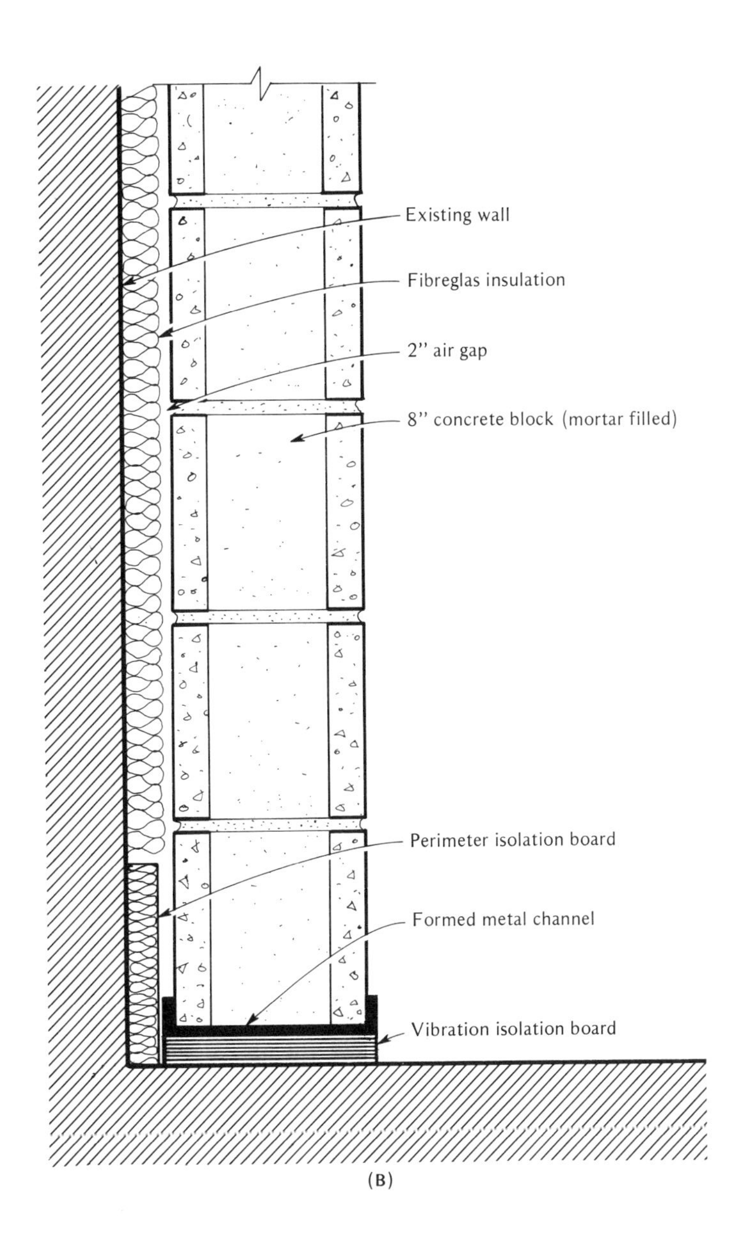

Figure 3-16 A typical floating wall (concrete)

floors

Wooden floating floors

Even when a room has tremendously insulative walls, sound and vibration can enter or leave by "flanked" paths through the structural floor. Thus, expensive sound-control construction can be easily "short-circuited" and rendered worthless. To prevent this, a resilient "floating" floor must be built.

A floating floor is simply a floor surface supported by a vibration-absorbing medium—such as springs, rubber, foam or a blanket of some type. An inexpensive floating floor can be constructed from plywood, placed over wooden battens (spacing strips) that are supported by a continuous layer of sound-deadening board.

Wooden floating floors improve the vibration and sound isolation by allowing the adjacent walls to act at their maximum acoustical efficiency. More efficient wooden floating floors utilize a heavier framework supported by specially designed *vibration isolators*, placed at regular intervals beneath the floor. Filling the air space between the support joists of a wooden floating floor with sand increases the mass of the floor substantially, thereby increasing its isolation performance. The sand should be *kiln-dried* to avoid the ill-effects of residue moisture, which could lead to warping or cracking at a future date.

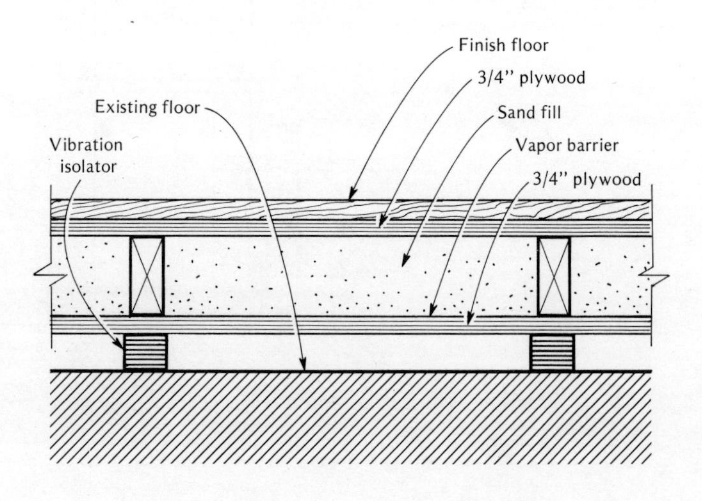

Figure 3-17 A typical floating wood floor

Care must always be taken during construction to see that the floating floor does not actually *touch* the walls surrounding it, thereby ruining the floating effect. This can be accomplished by lining the circumference of the room with *vibration-insolation board or rubber* before constructing the floor.

Concrete floating floors

A heavy concrete floating floor, working in conjunction with a substantial structural slab or sub-floor, can eliminate virtually all vertical transmission of noise through the studio floor. The basic method for building and floating a concrete floor is to support the floor at discrete points with very efficient vibration isolators, thus creating a sound-attenuating air space between the floating floor and the structural floor or sub-floor.

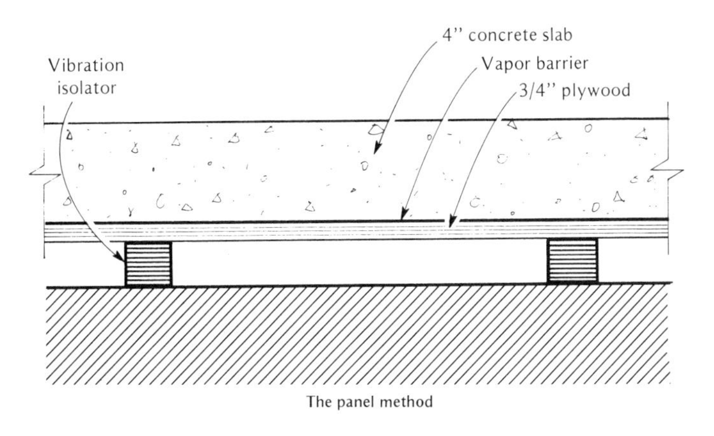

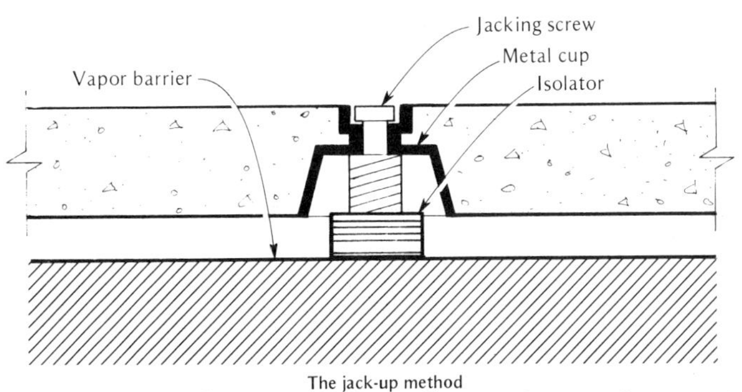

Figure 3-18 Floating concrete floors: The panel method vs. the jack-up method

In the *panel method*, the concrete floating floor is poured on top of plywood floor panels, which are in turn supported by vibration isolators. A thin plastic sheet is laid over the plywood to prevent the concrete from leaking through the spaces between panels. (See diagram for details). In the *jack-up method,* no formwork is used at all; the sub-floor is simply covered with a waterproofing layer of plastic, and the edge of the room is lined with asphalt-faced fiberglas. Then special jack-up vibration isolators are placed on the floor at regularly spaced intervals. The reinforcing bars are tied to those insulators, and the floor is poured.

When the floor has set, it is raised into position by turning the jacking screws in each of the vibration isolators. An adjustable air space from 1 to 4 inches is attainable. This method is illustrated in the accompanying diagram.

The jack-up method has several advantages over the panel method. The floor height can be finely adjusted to match the height of door thresholds, thus assuring smooth level transitions. Secondly, there are no chances of wet concrete breaking through the gaps in the plywood, thereby short-circuiting sound paths. Thirdly, there are no heavy formworks to deal with. Fourthly, the plywood panels in the panel method can be adversely affected by moisture and termites.

When vibration transmission through the floor is a severe problem, as in the case of a studio located near subway tracks, *spring-mounted* vibration isolators should be used. They have larger static deflections than regular vibration isolators (almost 1 inch, compared to the 1/4-inch deflection of regular isolators). These isolators can be incorporated easily into the jack-up assembly method (see accompanying diagram).

ceiling

A suspended or *floating* ceiling is an excellent device for attenuating the transmission of sound to and from above.

In order to insure the proper degree of sound isolation, a floating ceiling must be reasonably massive. (Acoustic tile will not work.) As a result, only a durable and rugged frame for the ceiling will be suitable. Normally, a float-

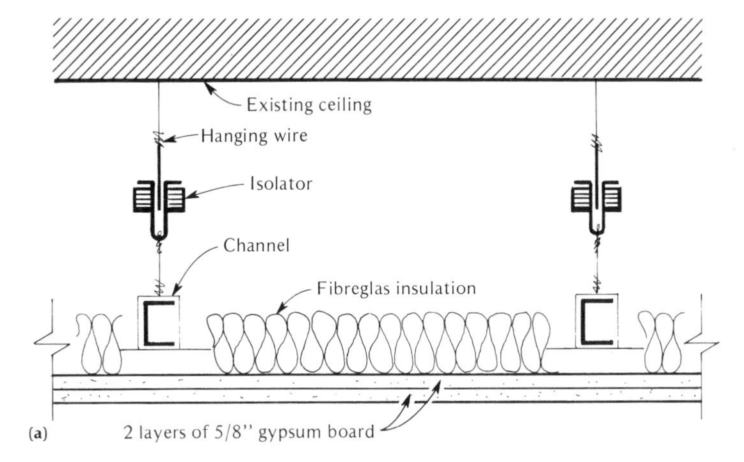

ing ceiling is hung from above, with *vibration isolators* made from neoprene rubber or coiled springs. This insures resiliency and, hence, maximum acoustical efficiency of the ceiling.

The periphery of the ceiling should be 1/4-inch clear of the surrounding walls, to insure that no rigid contact occurs. The resulting air space should then be caulked tightly with nonhardening caulking compound. The space above the ceiling should be lined with absorbent material.

Occasionally, a floating ceiling is supported from *below* by the floating walls. In this case, the weight of the ceiling is transferred directly through the walls, and special measures must be taken to insure that the resiliency of the wall system is not altered.

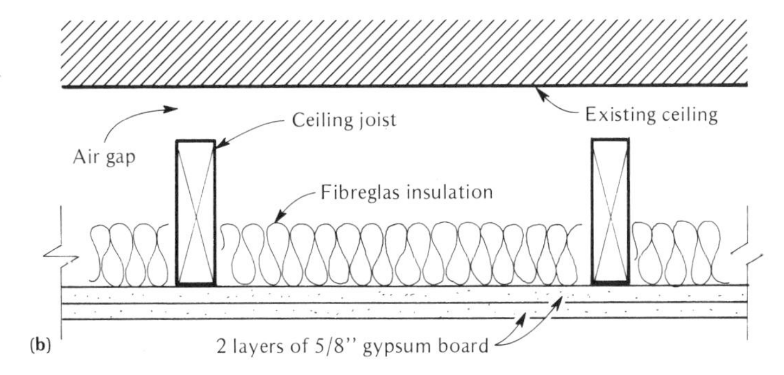

Figure 3-19 Floating ceilings suspended from a) above b) below

doors

Doors are the least massive, least airtight part of the floating studio's walls. Thus, they are the studio's weakest link in the sound-transmission chain. To alleviate this problem, each passageway leading out of a floating studio or control room must have at least two doors—one for the inner shell and one for the outer shell. This creates an air pocket between the doors called a sound lock.

The larger a sound lock is, the more usable it becomes. When properly designed, a sound lock can serve as a vocal booth or isolation room, useful in overdubbing. If properly planned, one central sound lock can take the place of many small ones by offering access to the control room and the outside environment, as well as to the studio.

Only solid doors should be used. Wooden doors are usually satisfactory, but if noise transmission is severe, laminated double-doors, lead- and foam-lined acoustical doors or concrete-filled metal doors can also be used. Gasketing is a must, and door-closing hardware is important to insure a tight fit. If the doors are outfitted with windows, only thick double panes should be used, and the air space between the panes should be lined with absorptive material.

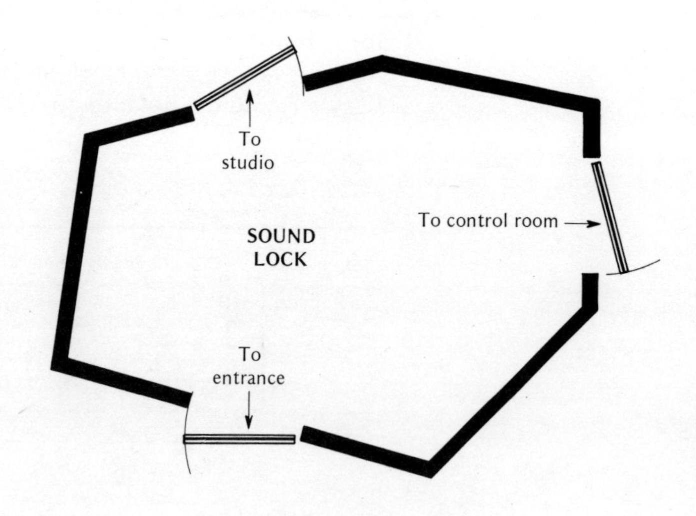

Figure 3-20 A typical sound lock

windows

Windows, like doors, are a weak link in the sound-transmission chain of the floating studio. For this reason, strict attention to detail in window construction is of paramount importance if the isolation of the studio is not to degenerate. Since it is almost always necessary to install at least one observation window in the studio for the purpose of visual communication between the engineer and the performers, we shall now turn to an examination of the proper construction of such a window.

The control room window

The control room window is one of the most difficult constructions in the studio. Visual criteria require that it be large (at least 10 square feet and sometimes as large as 200 square feet). Unfortunately, glass is a poor sound isolator. Therefore, special care must be taken so that the control room window does not degenerate the transmission characteristics of the studio. Obeying the *mass law*, glass is sound-insulative in proportion to its weight. However, in accordance with the floating principle, more than *three times* the predicted mass-law sound attenuation can be achieved with unconnected panes separated by an air space. Therefore, control room windows are constructed from two or more panes.

Best results are obtained when each pane is placed in an independent floating wall, so that sound vibration cannot be transferred from one pane to another by the adjacent structure. The glass panes should be skewed with respect to each other and opposing walls. This measure helps prevent the formation of flutter echoes and standing waves between the panes themselves, as well as between the panes and the walls of the studio or control room. It is advisable to use panes of different thicknesses in composite windows to avoid sympathetic resonance of the glass panels. As a final measure, the circumferential space between the panes should be lined with porous absorbent. For details, see the diagram on the following page.

These details are applicable to windows in isolation booths and partitions within the studios, as well.

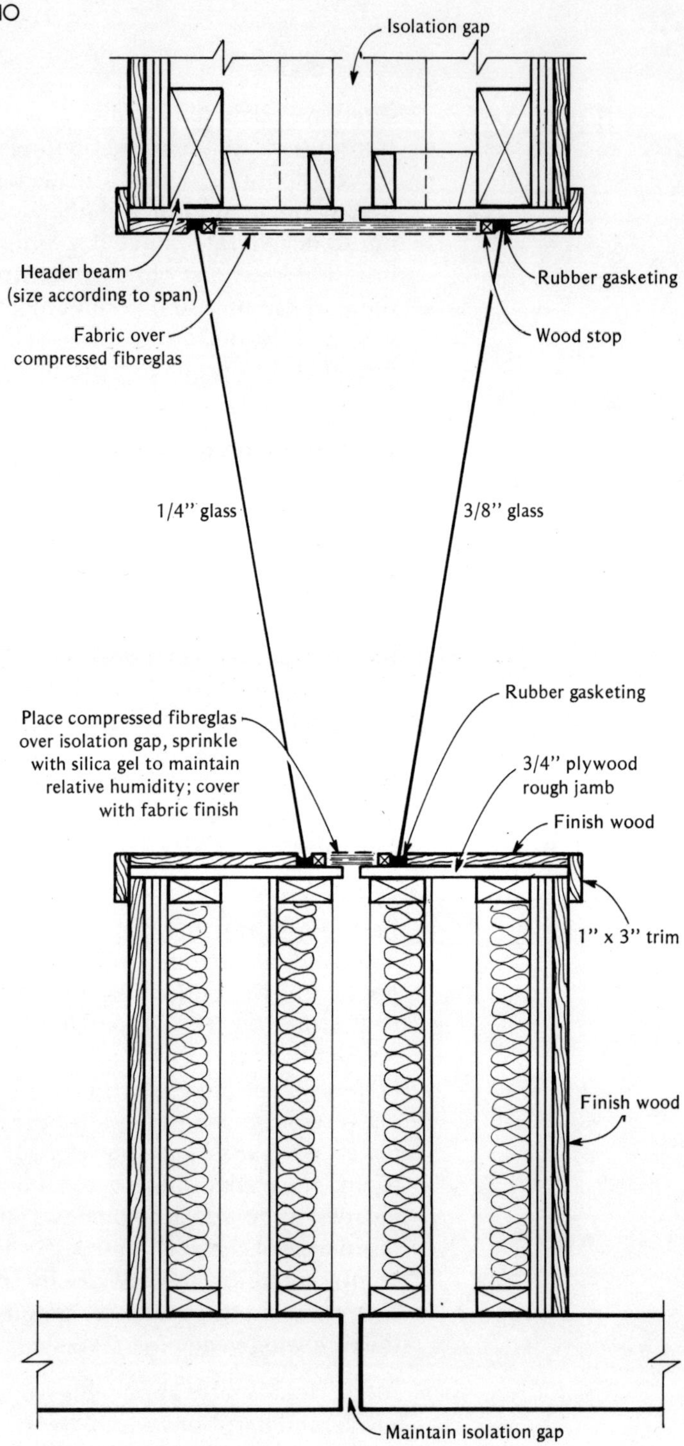

Isolation gap

Header beam
(size according to span)

Rubber gasketing

Fabric over
compressed fibreglas

Wood stop

1/4" glass

3/8" glass

Rubber gasketing

Place compressed fibreglas
over isolation gap, sprinkle
with silica gel to maintain
relative humidity; cover
with fabric finish

3/4" plywood
rough jamb

Finish wood

1" x 3" trim

Finish wood

Maintain isolation gap

Figure 3-21 Control room window detail

Other windows in the studio

Very often, it is desirable to incorporate windows other than the control room observation window into the design of the studio—for instance, if the studio is isolated in the countryside or overlooks a lake or mountain range, or if a little natural light is simply desired in the playing area. This type of amenity can be wonderfully advantageous to the look and *feel* of the studio, thus making a more comfortable environment to work and create in.

Traditionally, windows in recording studios have been discouraged, because designers feared isolation problems due to the inherent weak transmission values of glass. This fear would be justified if only a single pane or a double pane of glass were used. However, if a triple (or quadruple) pane is used and the details presented previously are closely adhered to, the isolation value of the window can be brought up to par with that of the surrounding walls and doors. Thus, if the expense of the procedure is not a deterrent and the surrounding environment is not prone to *excessively* loud noise (trains, trucks, jackhammers), then there is no reason to eliminate windows from the studio.

It should be noted that the presence of a large panel of reflective glass will influence the interior acoustics of the studio in a manner that might not always be desirable. For this reason, the window should be fitted with a thick curtain which can be drawn to increase acoustical flexibility.

ventilation

After every effort has been made to close off the studio from its external environment, the infiltration of fresh outside air will be almost completely impeded. Furthermore, the insulated air space that surrounds the inner shell will insure that there is very little, if any, heat loss from the studio environment. As a result, the heat created by lights, musicians and equipment will build relatively quickly.

Aside from being uncomfortable, this situation is positively unsafe.

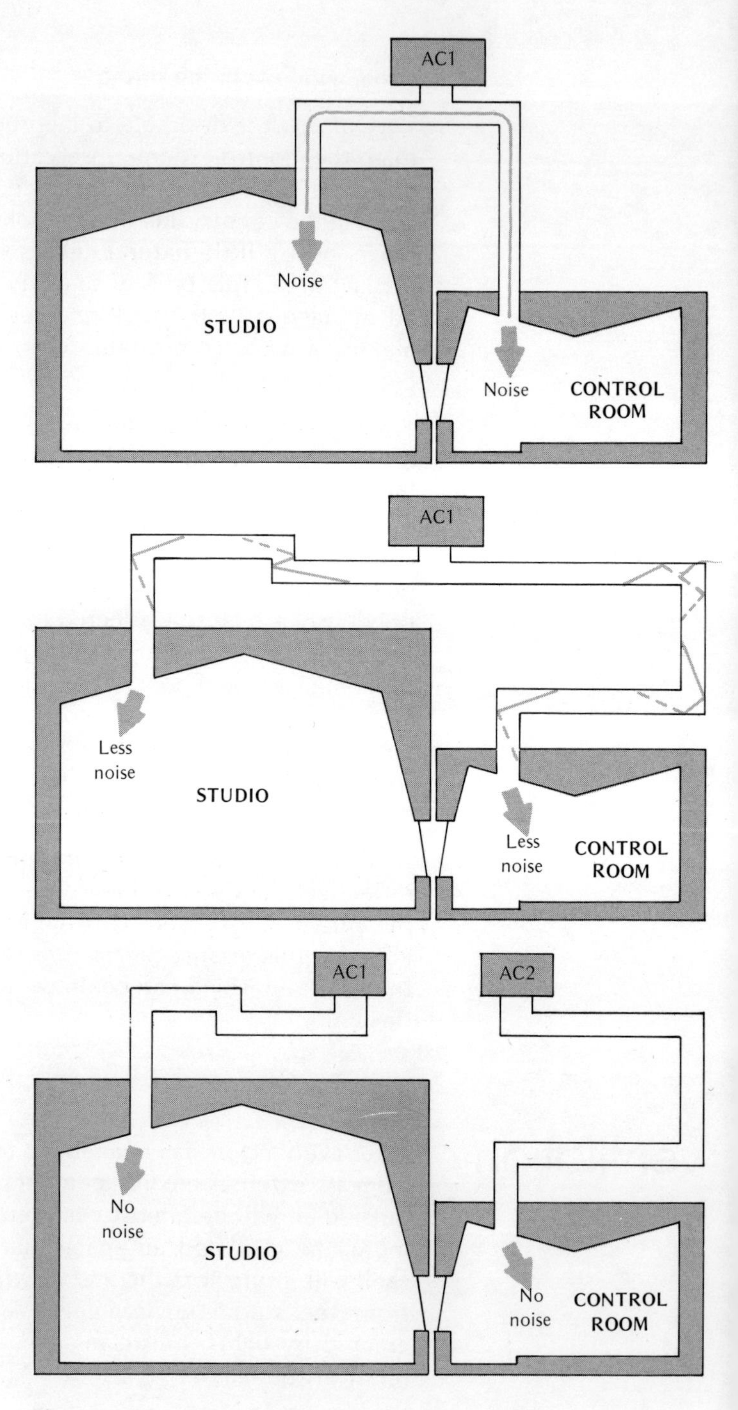

Figure 3-22 Sound transmission from studio to control room via ventilation ducts

Therefore, most studios must be ventilated and air-conditioned—despite the location. This holds true for almost all climates (even cold northern climates), but applies especially to temperate and sub-temperate zones (such as those in the U.S., Canada and Europe).

Unfortunately, air-handling equipment such as fans and compressors are large noise and vibration producers. Furthermore, the need for air to be delivered from the outside environment to the inside environment will inevitably involve penetrating the sound shell we have gone to such pains to construct, resulting in the free passage of noise from the outside to the inside environment and vice versa. This fact is a necessary evil of studio construction. It must be dealt with in a proper manner or all of the soundproofing effort spent in the studio will be short-circuited. A great deal of time and construction money will have been wasted.

Installing an isolated air-conditioning and ventilation system is a complicated matter. The problems that must be solved include:

1. Isolation of compressor and fan noise transmitted through the building structure.
2. Isolation of mechanical noises transmitted through the actual ductwork.
3. Elimination of air-movement noise through the duct and grille work.
4. Minimization of air-delivery velocity to eliminate interference with recording microphones.
5. Isolation of external noise transmitted through the ductwork.
6. Maintenance of the integrity of the acoustical shell while simultaneously penetrating it with ductwork and connecting it to the existing structure.
7. Controlling the distribution of air to the studio, control room, isolation booths and peripheral areas in a manner that provides for a very responsive control of temperature and humidity under a wide range of climactic conditions.

The solutions to these problems vary from studio to studio, depending on the construction, the mechanical equipment and the space available. When critical performance is needed, an acoustical consultant who specializes in this area should be engaged.

There are several rules that can be followed to minimize the problematic effects of heating, ventilating and air-handling equipment:

1. Compare noise and vibration specifications of equipment before purchase.
2. Locate all air-handling equipment as far away from the studio as possible. *A minimum of 30 feet prior to breaking the sound seal* is a good rule of thumb.
3. Locate equipment either on the ground or an *independent* slab. Avoid rigid mounting of the vibration-producing elements (compressors, fans) to the frame of the building. If this is not possible, mount the equipment resiliently on springs or rubber pads. Consult with a specialist for optimum spring-deflection values and inertia-pad construction.
4. Use *low air velocities* (delivery and return rates under 400 feet per minute) and large ducts. By reducing velocity of the air, "whistling" effects can be reduced or eliminated.
5. Introduce *90-degree bends* into the ductwork to create intentional baffling for airborne duct noise.
6. Line all ductwork with *sound-absorbing duct liner*.
7. Install "sound traps" at intermittent locations in the air-delivery system.

wiring

The need to interconnect electric devices has many important consequences in the construction and soundproofing of the studio. Primarily , it becomes necessary to puncture walls, floors and ceilings in order to locate conduits, panels and fixtures. These holes can seriously degrade the soundproofing of the studio.

To further complicate these matters, conduits and wiring chases are often constructed of rigid materials to comply with building codes. Thus, two isolation walls through which a conduit passes can inadvertently be tied together, thereby short-circuiting or ruining their effect. The solution to this type of problem can be rather intricate and often varies from instance to instance. In general, the following rules should be applied:

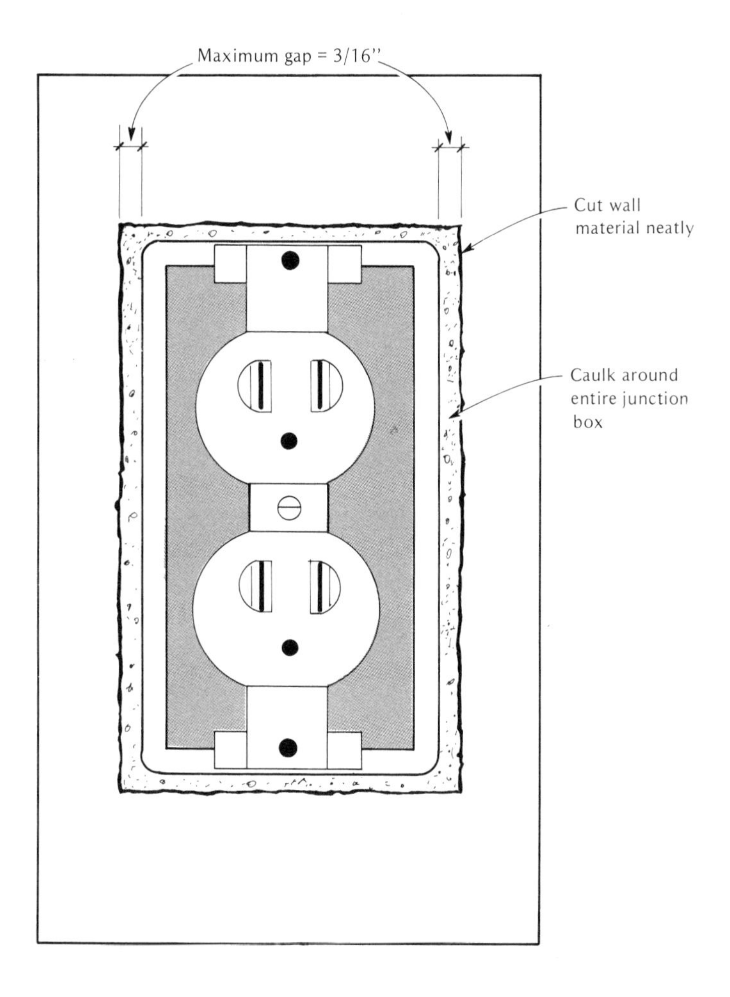

Figure 3-23 Caulking air gaps around electrical junction boxes to minimize sound leakage

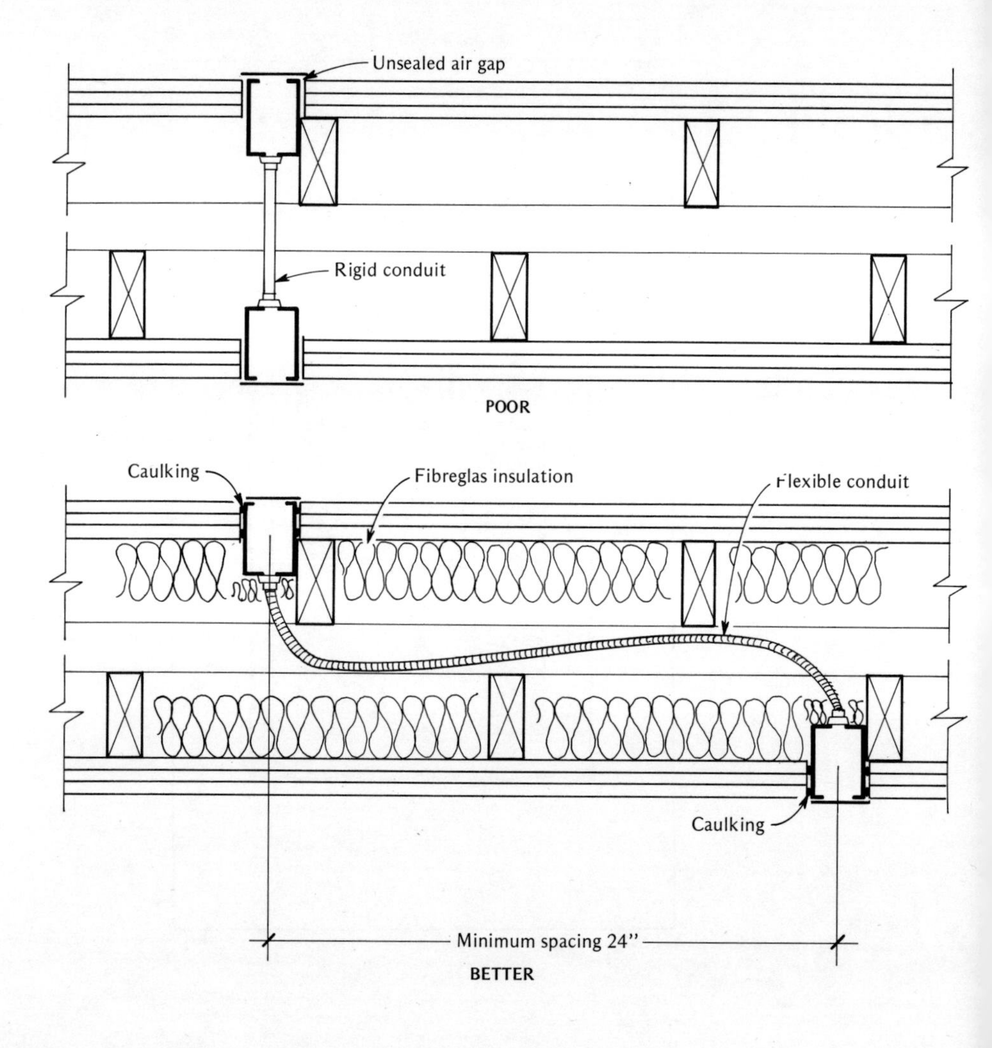

Figure 3-24 Minimizing sound leakage by proper spacing of electrical junction boxes

1. When crossing isolation lines, install an isolation "break" or gap in a rigid conduit to maintain integrity of the shell.
2. Build a recessed isolation box around the mike input panels and electrical junction boxes in order to eliminate the transmission of sound directly through the cavities created by these devices.
3. Any holes that must be made in adjacent walls should be staggered in order to minimize direct flanking of the sound path.
4. Caulk all air spaces with a non-hardening caulk compound.

chapter four

the studio

introduction

Once the noise problem has been solved by soundproofing, the task still remains to convert the room into a natural-sounding, comfortable environment for the playing and recording of music.

At the onset, this means eliminating standing waves, unnatural colorations, echoes, flutter, room resonances, focal points. These problems plague all rooms and are the first concern.

The next step is to optimize the studio for the particular type of music that will be recorded. If rock-and-roll will be the predominant music, isolation might be the most important criterion. On the other hand, if classical music or strings and horn music are to predominate, the natural ambience of the room might be the most important factor. When many types of music and instruments are to be accommodated, the acoustical qualities of the room must be versatile and adjustable to suit changing needs.

Last but not least, the studio should be designed to be an inherently warm and comfortable space. This will make musicians relax and undoubtedly improve the quality of their performance. The influence of the room aesthetics can often affect the caliber of the recording as much as the acoustical properties. Light, color, finishes and arrangement of space all play a role in determining the feel of the studio.

Combining technical performance with a beautiful and stimulating environment is the true challenge of building the ultimate studio.

location

The location and orientation of the studio can have a tremendous effect on the amount of soundproofing required. Ambient-noise measurements can differ substantially from one part of a building to another, depending on proximity to traffic, mechanical equipment, elevators and other noise sources.

In general, if the choice is available, the studio should be located as far from noise-producing elements as possible; also, the control room and storage spaces (where acoustical isolation is less critical) should be used as *buffer zones*. This will reduce the soundproofing that the stud walls must provide and therefore reduce the cost of construction.

Topography can also be used to advantage to reduce construction costs. Hills or ledges can act as "soundbreaks" if they occur directly in the path of oncoming sound waves.

The isolation effect is also true for high, solid walls such as masonry or stone built around the property. It is definitely *not* true for trees or shrubbery due to the myriad of air spaces through which sound waves can travel unimpeded. Contrary to a widely held notion, trees and shrubbery have little or no effect on acoustical isolation.

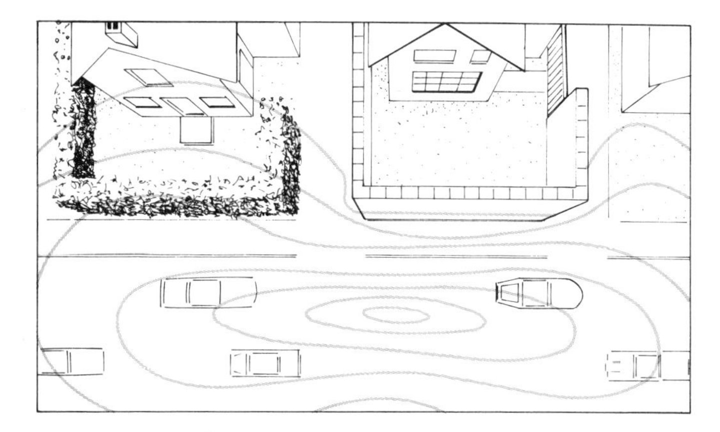

Figure 4-1 High walls can be used to acoustical advantage while trees and shrubbery have no effect

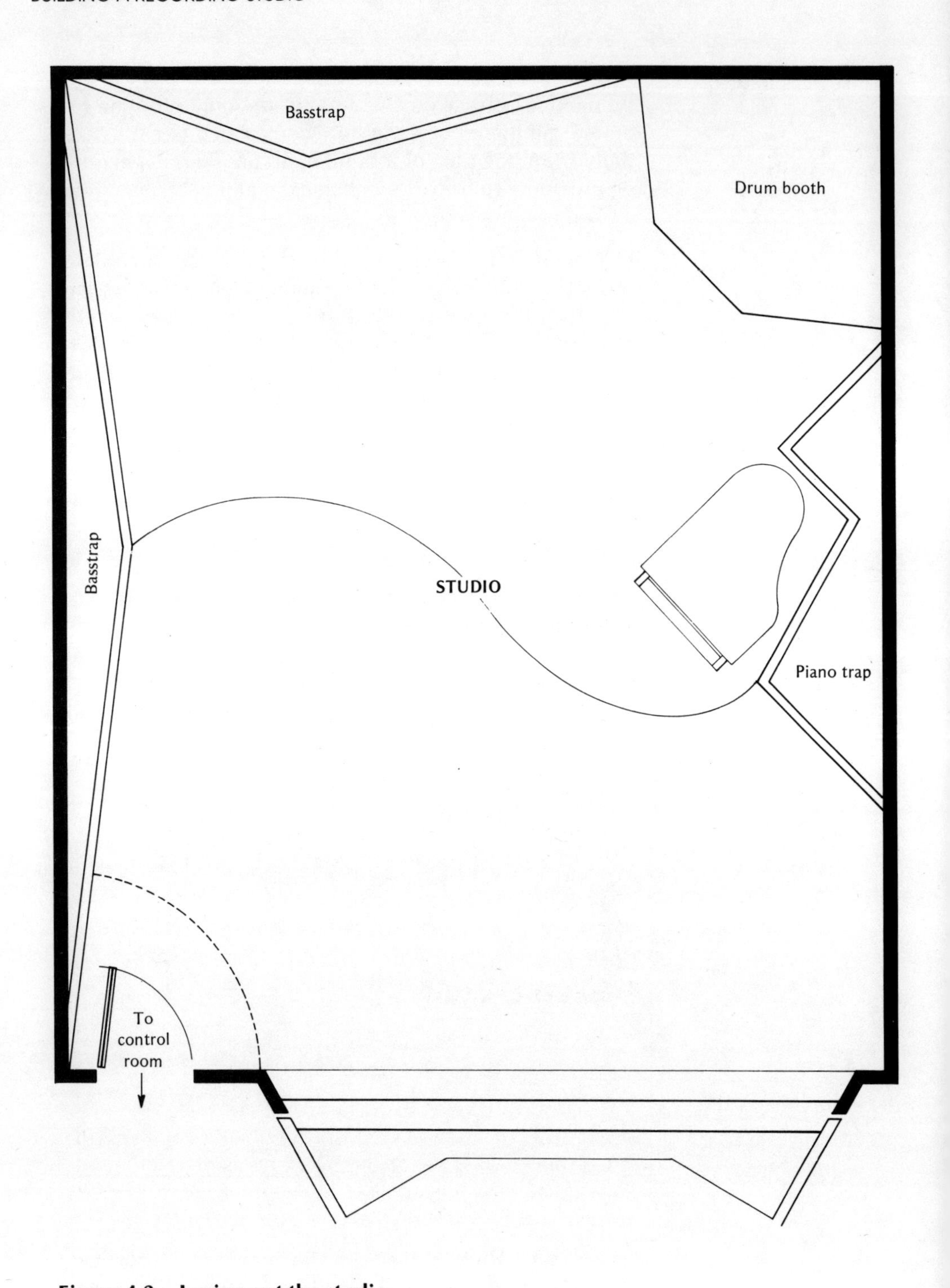

Figure 4-2 Laying out the studio

he floor plan

Laying out the floor plan of the studio is the most important preparation that can be made before construction commences. Space can be allocated, entrances and circulation patterns can be plotted, instruments and equipment locations can be determined, and construction costs can be estimated.

The floor plan should be drawn to scale (typically 1/4 inch = 1 foot or 1/2 inch = 1 foot) and show the following:

- control room location
- studio location
- the location of all new walls
- all entrances, exits and sound locks
- major equipment locations, including console, tape machines, equipment racks, etc.
- location of major instruments (piano, drums)
- electrical outlets
- air-conditioning outlets and inlets
- lighting fixture locations
- microphone panel locations
- basstraps and acoustical finishes

The construction details should also be noted. For this type of information, a sectional drawing or a large-scale drawing (3 inches = 1 foot) is often helpful.

Since each room has a unique size and character and each person's recording needs are an individual matter, there is no single studio layout that would be ideal for everyone—just as there is no single house design that is suitable for everyone. Each studio represents a design exercise, combining the requirements of the individual needs, the available space, the variety of uses to which the studio will be put, the aesthetic preferences of the owner and, not least of all, the budget available.

getting a permit

It is required by law to obtain a permit from the local building department before any major construction project or re-modeling job is undertaken. In general, the purpose of the permit is to allow the building department to inspect and insure the safety of all habitable construction with respect to fire, structural and other hazards. There is a permit fee payable which is often a function of the cost of work to be done. Local building departments will insist upon the right to inspect construction, to insure that all electrical, plumbing and carpentry work is performed in adherence to local building codes. Permits can be obtained by submitting plans for the work to the local officials of the building department or county engineers.

When applying for a permit to do work on a studio in a residential neighborhood, it is best to emphasize that the facility is only for private, and not for commercial, use. This should be stated clearly on the plans.

avoiding parallelism

Parallel surfaces are undesirable in studios, control rooms and listening rooms because they encourage the formation of echoes, flutter and standing waves. While the effects of parallel walls can be minimized by various means (see following sections), the best solution is to avoid parallelism in the first place. Unfortunately, most studios are constructed in rooms which already have parallel walls. The answer to this problem is to construct one or more new walls which are skewed slightly, so that no two opposing surfaces are exactly parallel. A skew of one foot for every ten feet of wall run (i.e., approximately six degrees), will be sufficient. Similarly, a skew of six inches for every ten feet of wall run for *each* of two opposing walls will accomplish the same effect.

Figure 4-3 Eliminating parallelism

Unfortunately, skewed walls do not discourage or eliminate *low-frequency* standing waves. These require special treatment described later in this chapter.

eliminating echoes

Distinct echoes occur when a sound wave bounces off a reflective surface in a room and returns to the listening area more than 35 milliseconds after the arrival of the original impulse. Since sound travels 1130 feet per second, this time interval represents approximately 40 feet of travel for an individual sound wave. This means that if a reflective surface in a studio is more than 20 feet away from a musical source, an echo is likely to be heard at the playing location and thus likely to be recorded (i.e., the sound will travel a total of 40 feet to and from the reflective surface).

Therefore, instruments which produce percussive wave forms (piano, drums, electric guitar) should not be farther than 20 feet from a large opposing reflective surface. If this situation should occur, the most effective solution is to cover the reflective surface with absorptive material, thereby reducing the echo-producing mid- and high-frequency reflections. If this is not possible (in the case of a control room window), the reflective surface should be *sloped* to focus the reflections away from the source and optimally towards an absorptive surface such as a carpet.

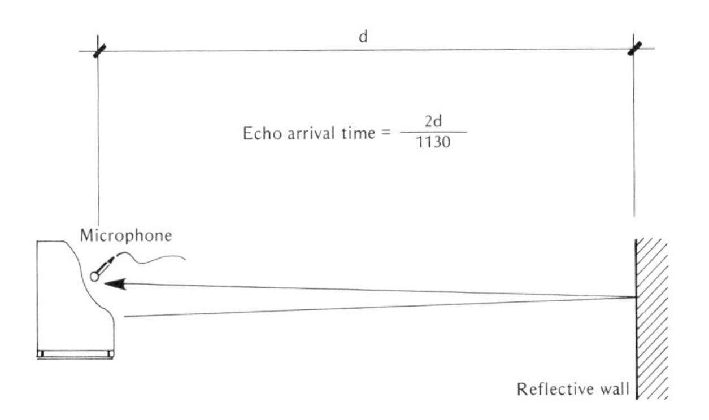

Figure 4-4 Echo from a distant surface

controlling flutter

Flutter echoes occur when sound waves bounce back and forth between hard, opposing parallel walls. There are four basic methods for eliminating flutter.

The simplest method to stop flutter is to cover *one* of the walls with a continuous layer of absorptive material. Another effective solution is to apply *patches* of absorptive material (e.g., tiles of cork or acoustic tiles) to *both* walls. The absorption deprives the reflected wave of some of its energy while the patching helps to eliminate flutter by causing the waves to diffract in random directions when they strike the patch boundaries. The patching solution helps maintain consistency between the character of the walls, whereas the first solution creates two walls of markedly different acoustical characters. Both have their own merits in particular applications.

Another solution to eliminating flutter, which leaves the reflectivity of both walls intact, is to slope one of the walls by ten degrees or more to discourage the direct reflection of waves back and forth. This solution is often used in echo chambers where absorptive materials cannot be applied.

A final method, which also leaves reflectivity intact and does not involve building a new wall, is to install large "diffusers" that cause the approaching sound waves to reflect in random directions. These diffusers often take the form of geometrical wooden constructions which protrude from the wall in a variety of shapes and sizes.

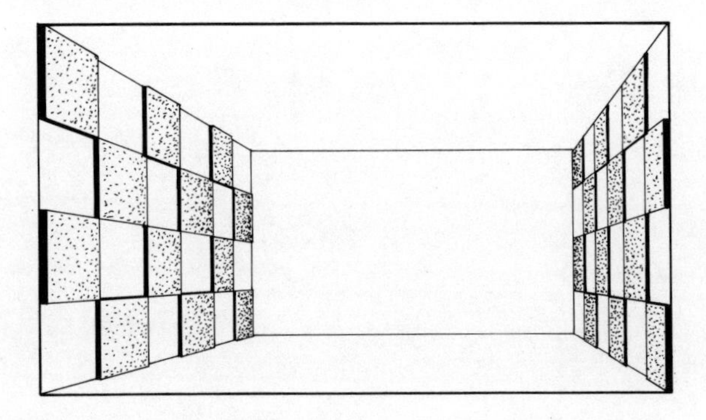

Figure 4-5 Controlling flutter by the application of patches of absorption

eliminating standing waves

Non-parallel walls, diffusers and absorptive patching cannot deprive standing waves of all their energy, especially in *small* rooms where standing waves form at very audible low frequencies. To help dissipate standing waves and the colorations they produce, special low-frequency absorbers must be built. These are called "basstraps."

There are many types of basstraps, some of which are quite intricate in design and construction. The most straightforward to build is the *membrane absorber*.

Membrane absorbers consist simply of a rugged membrane (such as plywood) stretched over an air space (see diagram). Each membrane absorber has a *fundamental frequency* determined by the weight and flexibility of its membrane material and the size of its air space. When a sound wave close to this frequency (or one of its harmonics) strikes the absorber, the membrane is set into

Figure 4-6 A typical basstrap

motion. This motion, in turn, attempts to set the air space behind it in motion. The resistance offered by the air space, combined with the "damping" motion of the membrane itself, helps dissipate and absorb the energy of the striking wave.

The effective frequency range of a membrane absorber can be increased by lining the air space inside with a porous absorber (fiberglas or acoustic tile). This tends to "flatten" out the absorption curve.

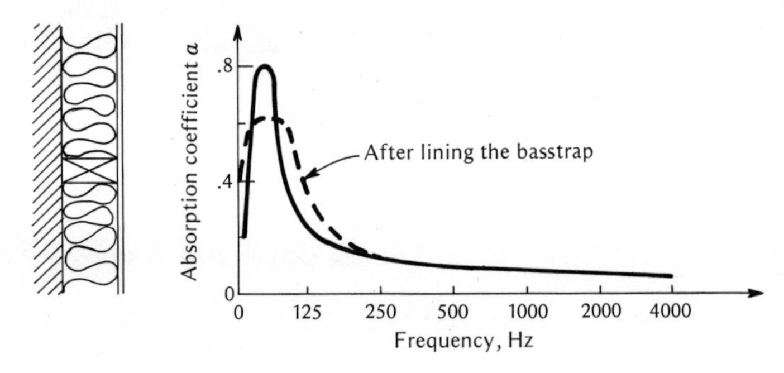

Figure 4-7 Effect of lining the basstrap

Each room will produce a complex set of standing waves, whose fundamental frequencies are determined by the room dimensions. Therefore, each room will require a unique variety of basstraps. Normally, the determination of these standing wave frequencies and the proper basstrapping necessary to eliminate them involves a series of mathematical calculations.

The following formula gives the approximate design criteria for *membrane* or *spaced panel absorber*.

$$d = \frac{28{,}900}{f_o^2 M}$$

where:

- d = depth of required airspace, inches (distance between back of panel and rigid wall)
- f_o = fundamental absorbing frequency of basstrap, Hz
- M = surface density of panel, lb./ft.^{2}

Note for plywood panels: *For typical ½" plywood, M = 1.375 lb./ft.^{2}. However, the weight of plywood can vary depending on exact composition and moisture content. To obtain the most accurate results from the formula, an actual 12" × 12" material sample should be weighed and the resulting value substituted in the formula for M.*

The following table has been designed to facilitate these calculations. It lists common room dimensions in five-foot increments and shows proper basstrap construction for each dimension.

Since standing waves can form in any direction, each dimension of a room should be noted (i.e., length, width and height). If the room has divisions or alcoves, these dimensions should be on hand as well. Standing waves can form easily in these areas.

BASSTRAP CONSTRUCTION CHART

ROOM DIMENSION (feet)	STANDING WAVES (Hz) (harmonics)		
	1st	2nd	3rd
5	113	226	339
10	57	113	170
15	38	75	113
20	28	57	85
25	23	45	68
30	18.8	38	55

ROOM DIMENSION (feet)	AIR SPACE REQUIRED (inches) THICKNESS OF PLYWOOD PANEL			
	1/8″	1/4″	1/2″	3/4″
5	6½″	3¼″	1¾″	1⅛″
10	25¾″	13″	6½″	4¼″
15	58¼″	29″	14½″	9¾″
20	107¼″	53½″	26¾″	(17¾″)
25		79½″	(39¾″)	(26½″)
30			(59½″)	(39¾″)

Figure 4-8 Basstrap construction chart

How to use the table

Locate each room dimension in the table above. Determine which standing waves are forming in that direction by using the formula on page 50. Then determine what combination of airspace and panel thickness are most practical for your particular room. In general, the thinner the membrane used, the more efficient and finely tuned the result will be.

Of course, most actual room dimensions will fall between the five-foot increments given in the table. Approximate values can be obtained by interpolation. However, to obtain more exact values, the original formulae must be used.

building the basstrap

The method for constructing and installing basstrapping is outlined below:

1. Determine the type of membrane that will be used, and check the required air space for your particular room in the preceding table. Do this for all dimensions of the room (length, width and height).

2. Apply nailing strips (or nailing assemblies) of the same dimensions as the required air space to the existing wall. (This can be the original wall or the secondary, sound-insulating wall built in accordance with Chapter III.) The strips should be spaced a minimum of 24 inches apart.

3. If a thick, heavy membrane is used (e.g., half-inch plywood or thicker), and the required air space is *small* (under 4 inches)—*single studs* of wood (such as "two-by-fours") may be used for nailing strips. (Note: the actual dimensions of a "two-by-four" are only 1½ inches by 3½ inches.) Since the air space dimension of a membrane is relatively critical, it is important to use wooden members of the correct size to achieve the exact air space required. If the required air space is *larger* (over 8 inches), a *nailing-block assembly* (made of several studs) will have to be constructed. Both of these are pictured.

4. Line the air space with fiberglas, acoustic tile or an absorbent blanket. This broadens the effective frequency range of the basstrap and helps to prevent audible resonances inside the air space.

5. Apply the membrane to the nailing assembly by means of direct nailing, screws or glue.

6. Seal any air spaces between the panels with caulking or cover the air spaces with moulding. The panel should be airtight.

The surface area that the basstraps must cover in a studio will depend upon many factors, including the size of the studio, the severity of its standing waves, the effectiveness of the basstraps, etc. Usually, the area to be covered will vary between 25% and 75% of the surface area in the room. *As a rule of thumb, if half of the wall and ceiling area is treated with resonators of the proper dimension—concentrated especially at corners and on*

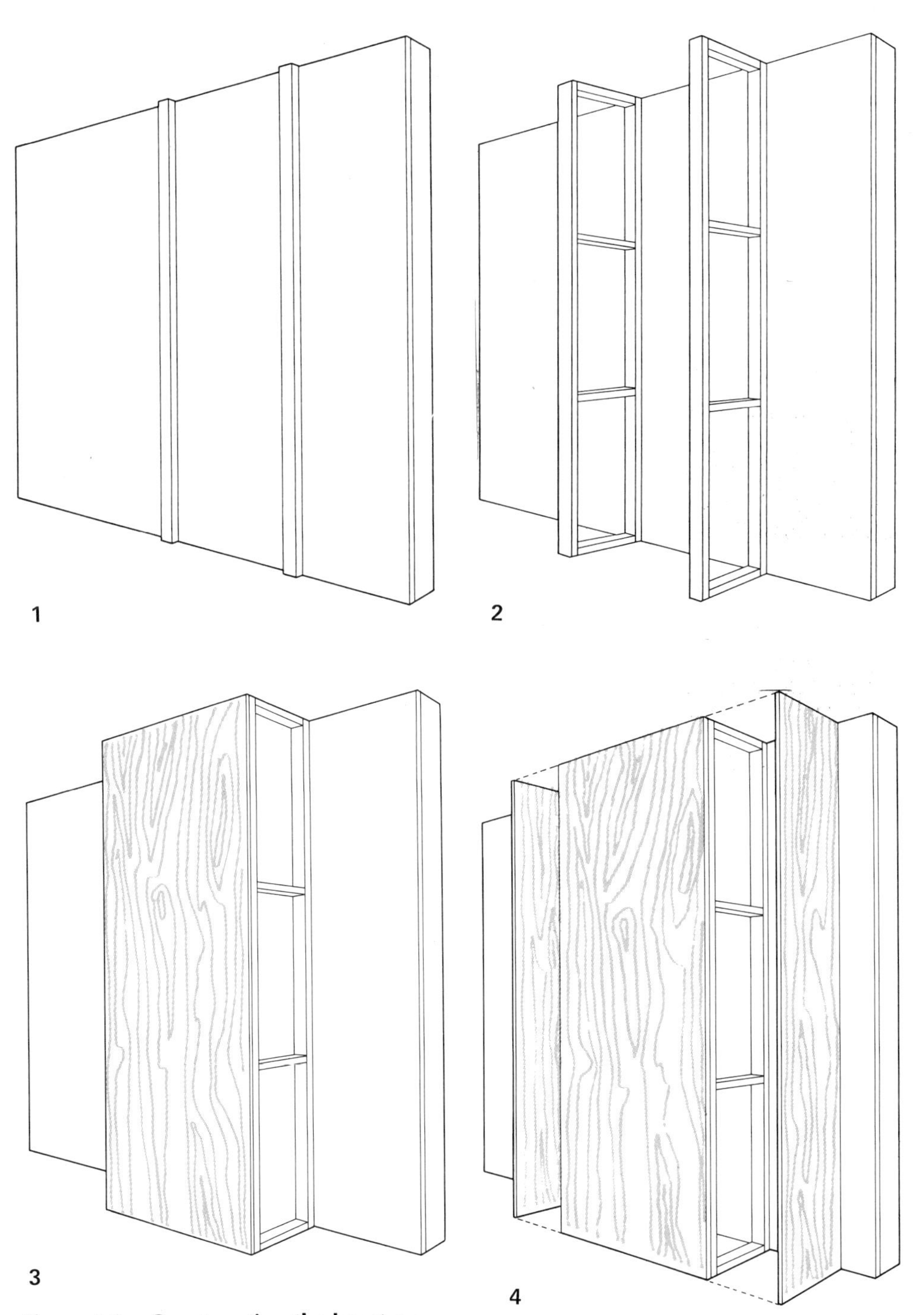

Figure 4-9 Constructing the basstrap

opposing, parallel walls—a studio will be reasonably safe from severe standing waves.

In general, smaller studios (under 150 square feet) will require slightly larger percentages of low-frequency absorption since their dimensions encourage more severe standing waves. Larger studios will need less basstrapping, proportionately speaking.

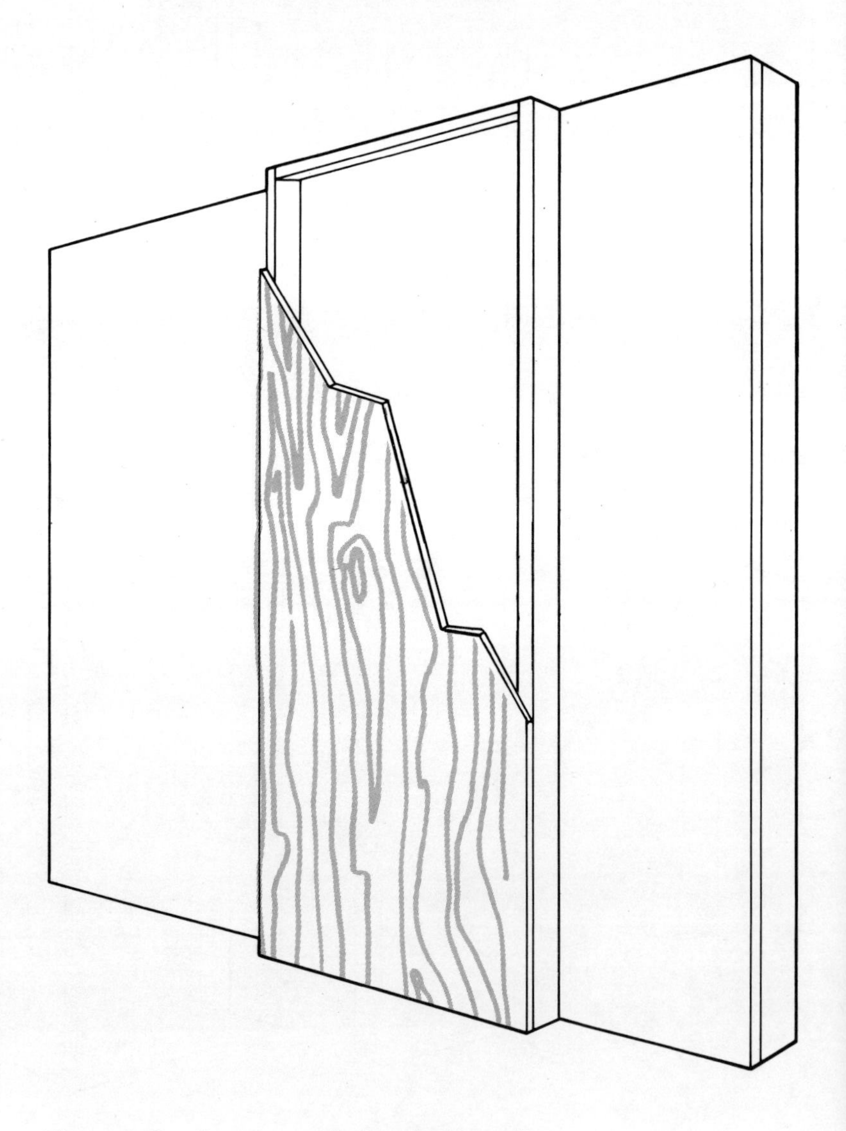

Figure 4-10 Basstrap utilizing single studs

the multi-dimensional basstrap

To broaden the frequency range over which the basstrap is effective, the nailing assembly can be skewed to create a multi-dimensional air space.

This type of assembly serves four purposes:

1. It absorbs low-frequency energy over a broad spectrum.
2. It helps break up standing waves, echoes and flutter by creating a non-parallel wall surface.
3. It helps promote the diffusion (even distribution) of sound fields in the studio.
4. It eliminates the need for exact calculation of basstrap dimensions, by providing a general-purpose low-frequency absorber.

As a result, this skewed type of basstrap is highly recommended for home studios.

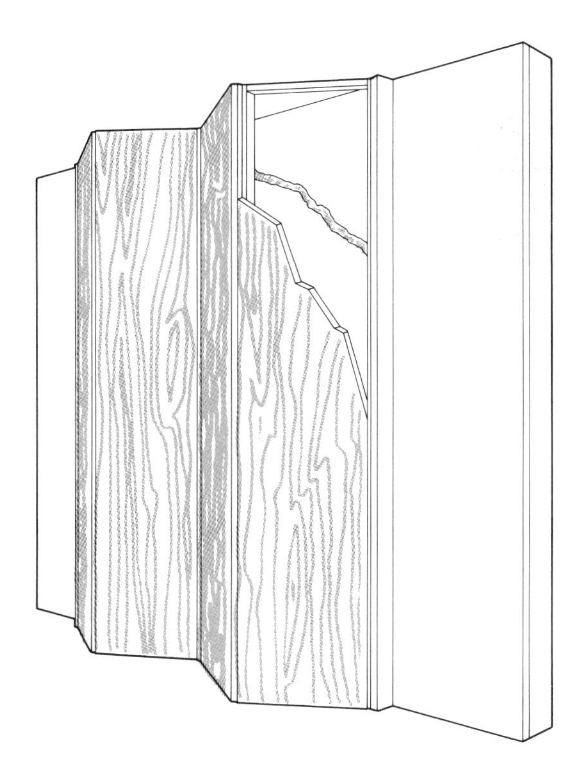

Figure 4-11 A multidimensional basstrap

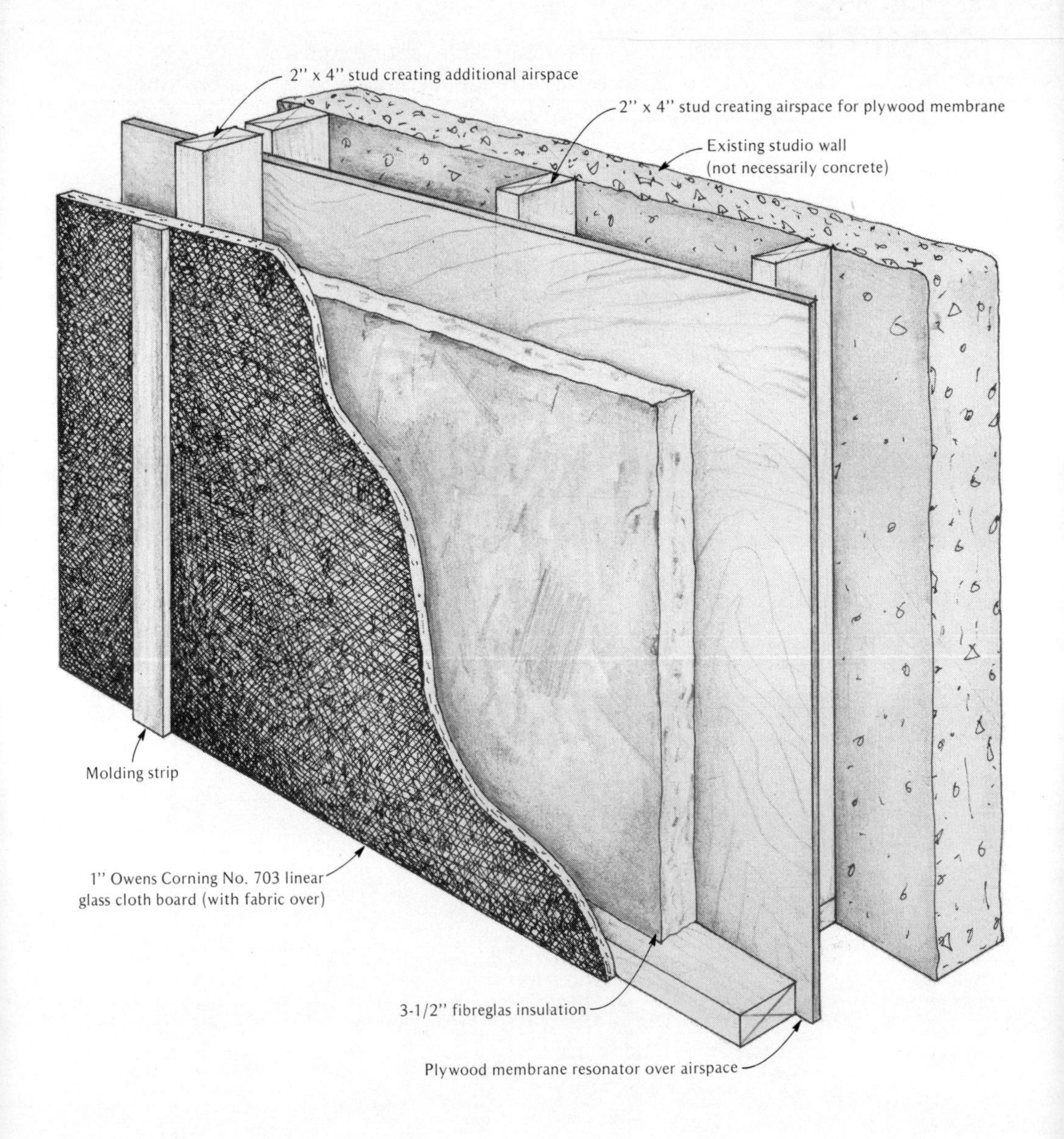

Figure 4-12 Facing the basstrap to create a broadband absorber

augmenting the basstrap

It is often desirable to make the basstrap absorptive at *high* and *mid* frequencies, as well as low frequencies.

This is accomplished by facing the basstrap with *linear glass board* (a relatively sturdy high/mid-frequency absorber constructed from compressed fiberglas).

The linear glass board is supported by wooden spacers at regular intervals. The resulting air space between the membrane and the linear glass board allows the membrane to resonate freely. If the linear glass board were applied *directly* to the membrane, it would change the panel's stiffness, possibly altering it's fundamental frequency. However, when the proper air space is maintained, the effectiveness of the basstrap is *not reduced at all,* owing to the ability of the low frequencies to pass directly through the linear glass board.

Broadband absorption can be increased even further if the air space between the linear glass board and the membrane is lined with a blanket of fiberglas insulation. The insulation, however, should not come in contact with the membrane for the reasons discussed.

This assembly creates a very effective *broadband absorber,* which has many applications in the studio and control room.

eliminating severe colorations

Most room colorations can be sufficiently dissipated or eliminated by the membrane resonators or basstraps discussed earlier. However, if unusual room configurations are producing objectionable colorations or standing waves at a particular frequency, then more specific remedies are needed. (The presence of severe colorations can sometimes be determined by a well-trained ear, but more often it is necessary to do audio testing of the room acoustics).

One of the most effective ways of eliminating a specific room resonance is to use a Helmholtz resonator. This device consists simply of an enclosed air space whose natural frequency is excited (or set into resonant motion) by the particular room resonance. The vibrating air space proceeds to dampen the room resonance by the process of phase cancellation, as discussed in Chapter II.

As a result, when we design a Helmholtz resonator for the studio, we attempt to create an air column whose natural frequency is the same as the room's standing wave or coloration. This column size is determined mathematically—and the resulting resonator is built in accordance with the calculated result.

The detection of severe standing waves and the determination of their exact remedies is a demanding process, involving testing and mathematical calculation, and usually requires a trained expert.

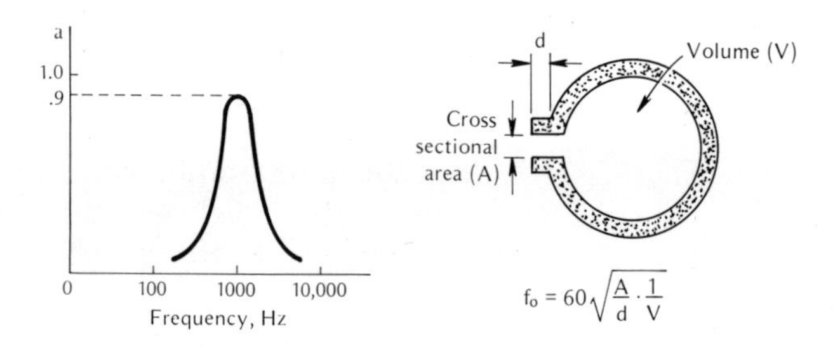

Figure 4-13 A Helmholtz resonator

gobos

During a multitrack recording session (in which individual instruments are assigned to separate tracks of the tape), the musical instruments must be *acoustically isolated* to prevent the sound of one from "spilling over" to the microphone of another. This goal is accomplished by three basic methods:

1. Designing the studio to discourage unwanted reflections.
2. Miking each instrument at close range with unidirectional microphones, which reject extraneous noise.
3. *Enclosing each instrument with portable sound isolators or "gobos."* These panels form makeshift alcoves around the individual instruments and help to locally isolate their sound waves.

To effectively retard sound at lower frequencies, gobos must be reasonably massive. This can be accomplished by constructing the gobo from plywood, particle board or gypsum board. Several sheets of these materials, applied in a sandwich construction, are most effective. A gobo constructed simply from fiberglas or blankets stuffed into a frame will *not* be nearly as effective as a solid gobo.

Unfortunately, the massiveness of the gobo does not entirely insure its ability to reduce sound transmission. Extremely low-frequency waves, which are quite large, can easily diffract around any small barrier, regardless of its weight. For example, a single gobo measuring 3 feet by 4 feet would do little to impede the sounds of an amplified bass guitar, whose wavelengths can measure over 25 feet long.

Therefore, the *physical size* of the gobo plays an important role in determining the lowest frequency at which isolation will be effective. Since gobos as large as 20 or 30 feet long would be impossible to use, several gobos are often grouped around an individual instrument, creating, in effect, a single large barrier.

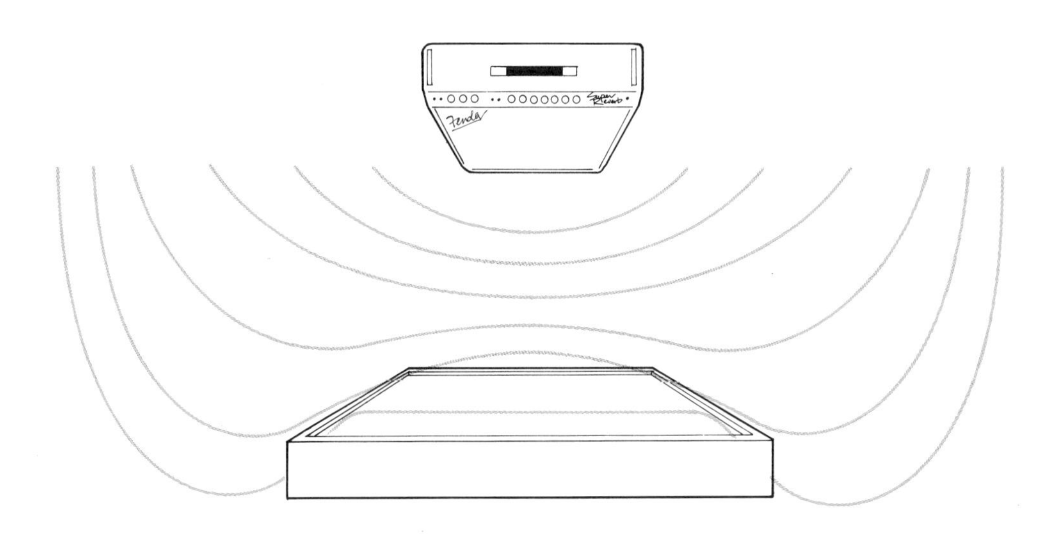

Figure 4-14 Low frequency waves diffracting around a gobo

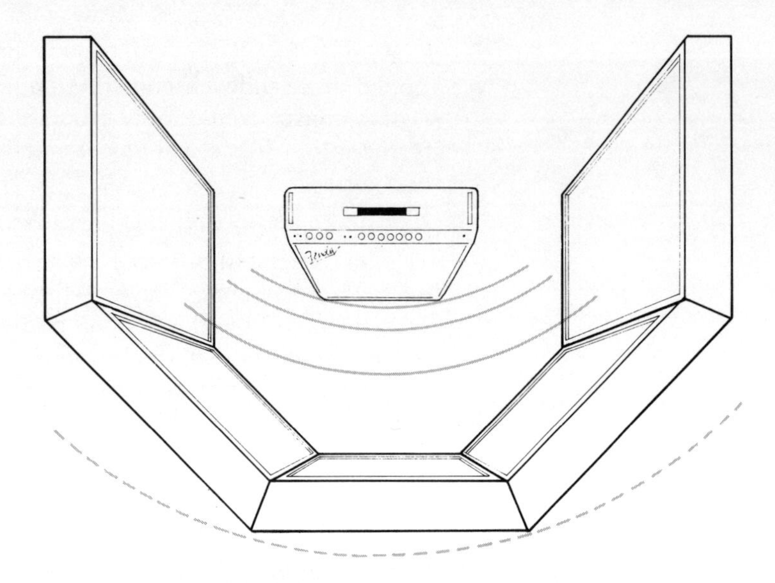

Figure 4-15 Containment of sound leakage

The surfaces of the gobos should be treated with absorbent material such as fiberglas. This will help to stop impinging sound waves from being reflected out into the studio (as well as back into the mike where they might cause acoustic phase cancellation).

This type of absorbent surfacing gives the instrument being recorded a rather "dry" or isolated sound. Sometimes, this is desirable (especially, if effects such as reverberation or echo will be added to the instrument during mixdown). If, however, a more "live" room sound is desired, only *one side* of the sound baffle should be lined with fiberglas. The other side should be left *hard,* thereby offering a reflective surface to the sound wave. This hard surface (e.g., plywood) can be painted or stained so that it looks attractive. This double-sided gobo can then be turned around during the recording session to give the sound that is most appropriate.

There are many satisfactory designs for gobos. One that has proven both effective and easy to build is shown in the diagram on the following page.

In order that a sound baffle can be easily moved by one person, casters or rolling wheels are often installed on

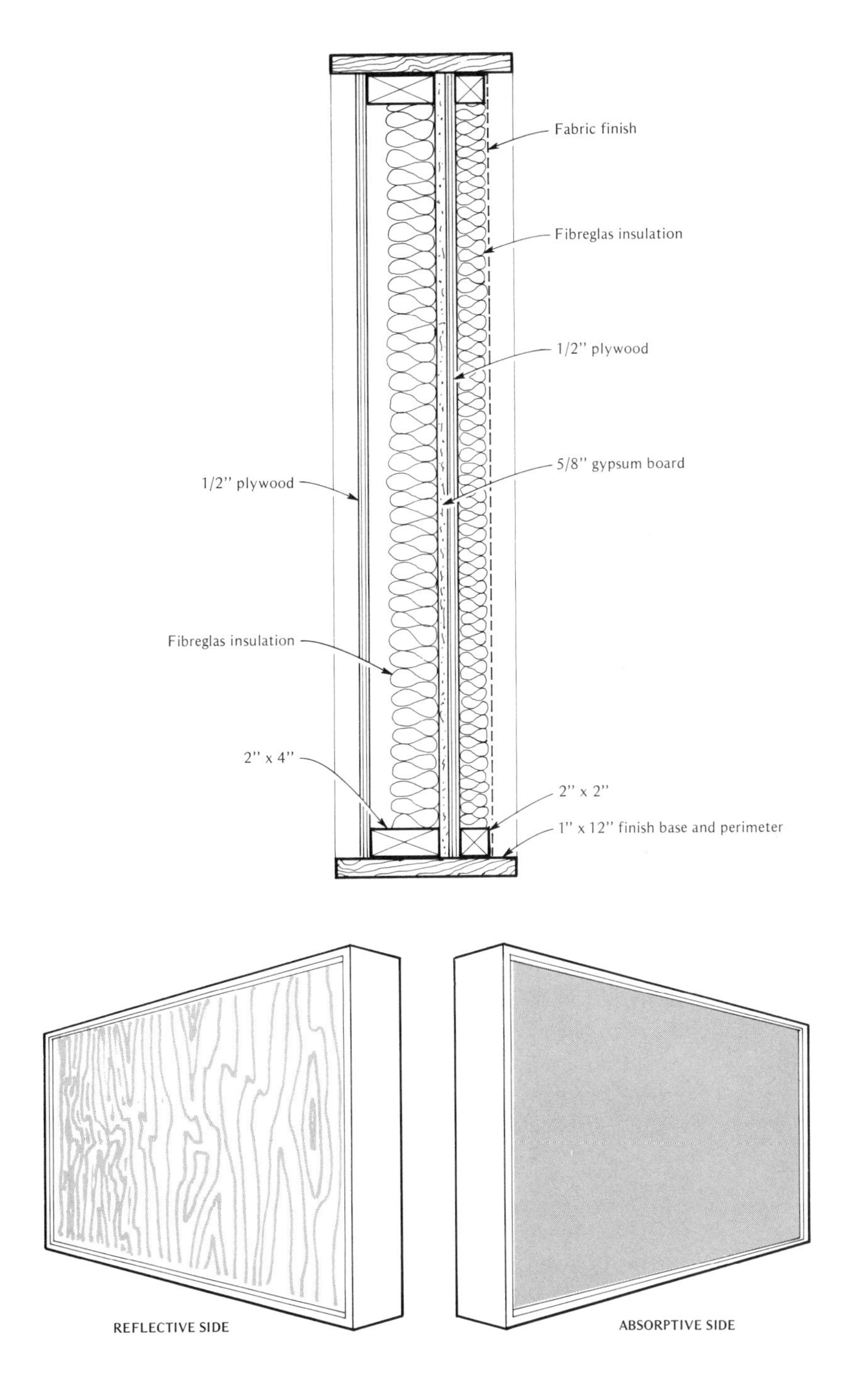

Figure 4-16 Satisfactory design for gobos

the base. Unfortunately, this can create an undesirable air space underneath the gobo, which can leak tremendous amounts of sound. This problem can be solved by lining the bottom edge of the gobo with a continuous double *skirt*, made from strips of rubber or carpeting.

drum booths

An aggressive drummer can create a tremendous amount of sound energy (110 dB (spl) or more). This can cause severe leakage problems in the studio. The first effect is that unwanted drum sounds find their way into microphones meant for other instruments. This complicates mixing and makes achieving the desired sound on the individual tracks almost impossible. A second effect, however, is sometimes even more disturbing: Because the drum leakage picked up by distant microphones is out of phase and delayed in comparison to the actual drum sound, the resulting *recorded* drum sound becomes loose and muddy-sounding when all the tracks are combined.

The solution to these problems is to incorporate into the studio some means for isolating the drums, such as a "booth" or "drum area." To be successful, the drum booth must first contain an isolated *platform* so that vibrations from the kick drum will not find their way into every microphone stand in the studio. The platform must be massive, not only to provide proper isolation but to help create a solid, thudding bass drum sound. This massiveness is accomplished by filling the platform with sand or concrete. Loose platforms constructed of a single layer of plywood will inadvertently act as membrane absorbers, thus depleting the low frequencies and weakening the drum sound.

The platform should be floated on a vibration-isolated medium such as fiberglas, celotex or prefabricated isolators to decouple it from the rest of the studio.

The booth should have an overhead canopy to absorb sound and prevent miscellaneous reflections from entering the studio. The lower the canopy, the better the isolation will be. However, absorptive canopies that are too

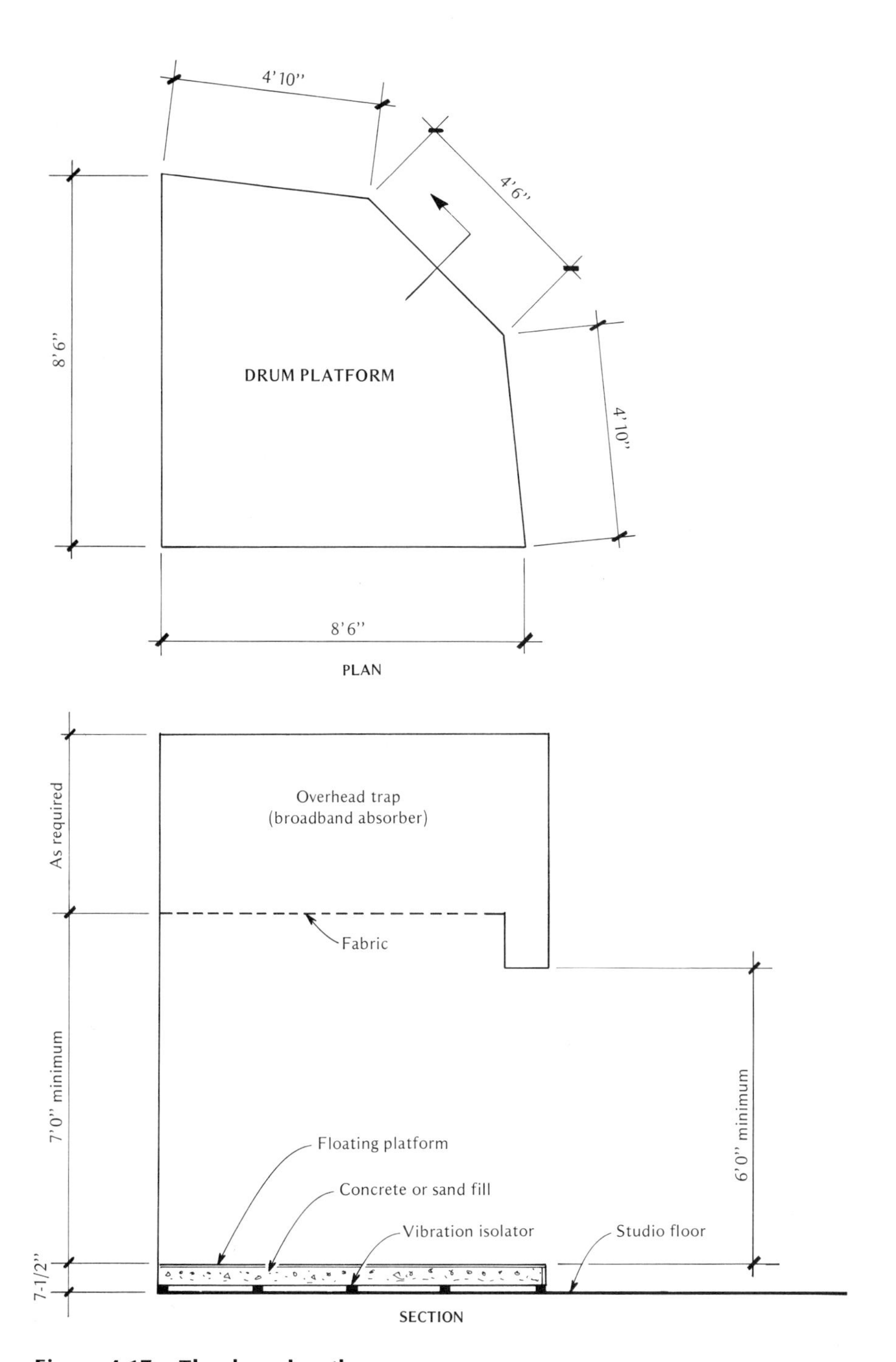

Figure 4-17 The drum booth

low (under seven feet) can be oppressive to drummers and deaden the drum sound.

Basstrapping is often applied above the drummer's head, in the canopy interior, to absorb unwanted resonances and overhead reflections. The amount of absorption required varies with personal taste—there is no general rule.

The booth is often placed in the corner of the studio, and the adjacent walls are treated absorptively to prevent high-frequency reflections produced by the cymbals and high hat from splashing into the studio. Gobos can be placed around the booth or incorporated permanently into the booth design to increase isolation.

live vs. dead areas

Since a studio is often required to suit many types of musical instruments, it is often necessary to incorporate some *acoustical flexibility* into the design of the studio.

For large studios (over 1000 square feet), this may be accomplished by allocating specific "live" and "dead" areas. The *live* area will be characterized by *hard* reflective surfaces in the walls, ceiling and floor, producing lengthy reverberation time and room ambience. This optimizes the sound of strings, horns, woodwinds and other orchestral instruments which require blending and reinforcement.

This live area would normally be located close to the control room window to take advantage of the reflective characteristics of this surface. It is important to note that although reverberation is encouraged in the live area, a properly designed live area is *not* simply an area with 100% hard surfaces. If the live area is not acoustically treated, it will suffer from severe coloration problems (i.e., the natural sound of the instruments will be altered). In addition, *leakage* between instruments at low frequencies will be excessive, causing a muddy-sounding recording. The key to the solution is to incorporate properly designed *basstraps* into the live area. This alleviates coloration and eliminates low-frequency leakage. The traps themselves must be reflective at the high frequencies to maintain the critical ambience in this part of the spectrum.

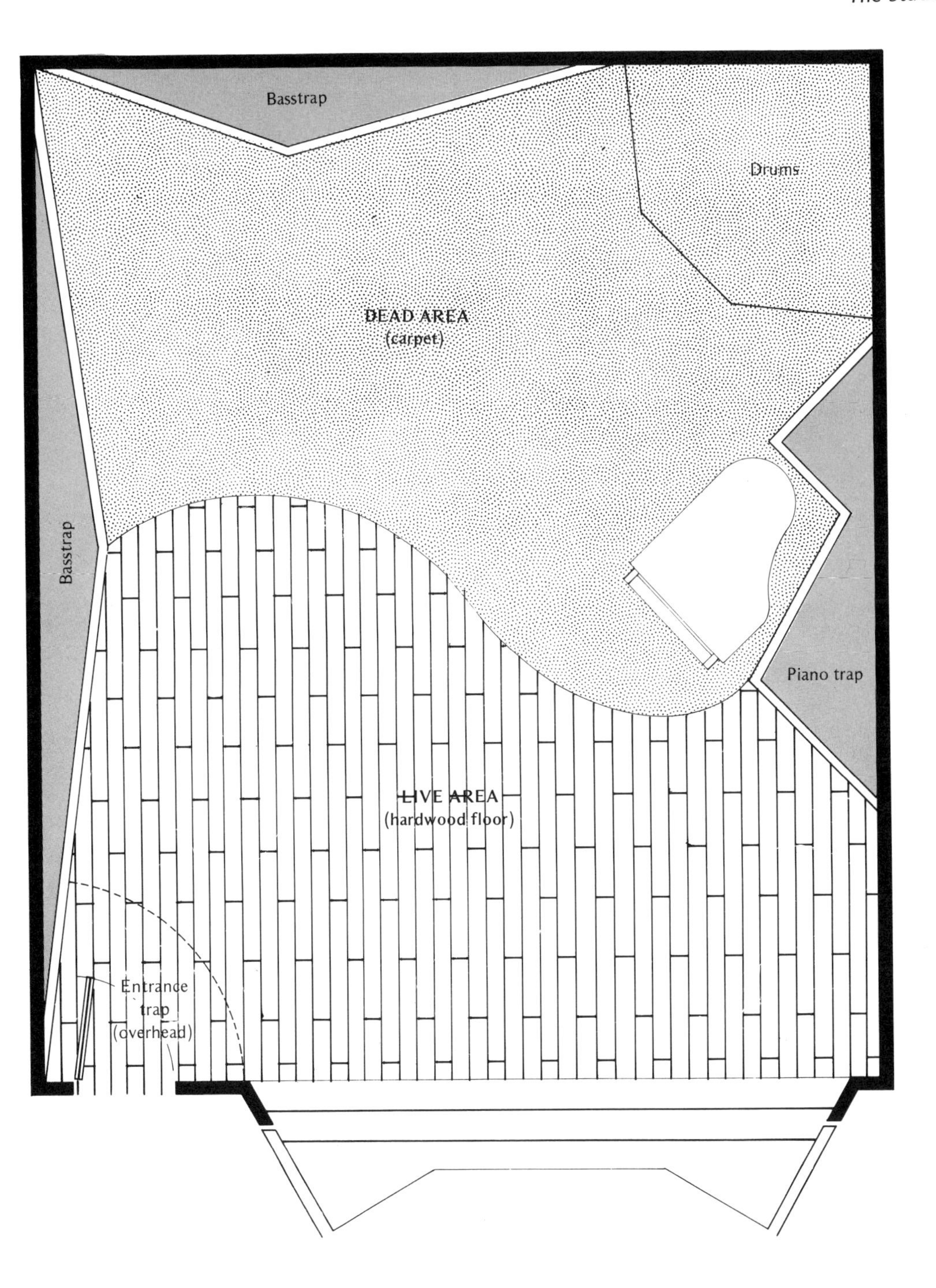

Figure 4-18 Live vs. dead areas in the studio

The *dead* area on the other hand, would contain many absorptive surfaces which prevent the successive reflection of sound waves. This enables instruments that produce relatively high spl levels (drums, electric guitar, bass) to be recorded with a high degree of isolation. It is important to note that making an area truly dead does not just mean applying carpeting and curtains. These materials absorb mainly high-frequency sounds and as a result do not reduce leakage at low-frequency levels. The dead area of a studio must contain ample *basstrapping* in addition to high-frequency absorption to insure isolation at all frequencies.

acoustical flexibility

If the studio is not large enough to divide into separate live and dead areas, another approach may be used. In this method, *variable absorbers* are applied to the walls to enable the acoustical climate to be changed upon demand. Variable absorbers can take the form of sliding wall panels, which deaden the sound of the room when slid into place. Portable sound baffles, which can be located as desired, can also be used to provide needed absorptivity.

Hinged wall panels with reflective and absorptive sides are quite effective for varying the acoustics of the studio. These panels can be used to transform the acoustics of a studio through varying degrees of "liveness," as well as to form mini-enclosures or booths for solo instruments. These functions are illustrated in the accompanying diagrams.

The panels can be constructed in a similar manner to the gobos discussed in the previous section. (See gobo diagram.) In fact, the panels act as vertical, wall-hung gobos.

Curtains and removable rugs may also be used as temporary absorbers. To be effective, a curtain should be as thick as possible, with a strong fabric backing. The actual curtain should be made from sound-absorbing material such as heavy velour or flannel. Light chiffon, gauze or bamboo curtains do not absorb sound. Better low-frequency absorption will result if an air space of

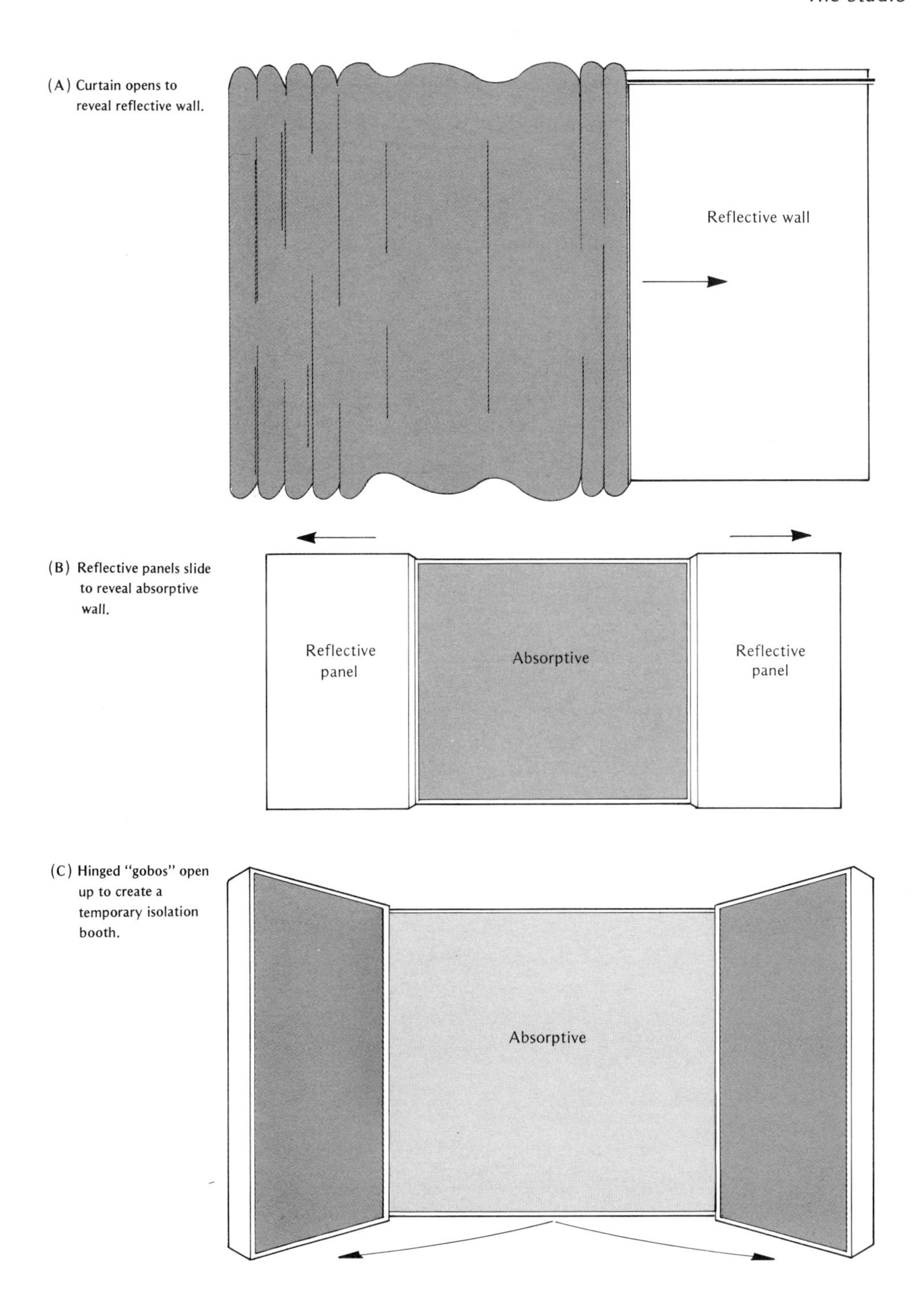

Figure 4-19 Variable absorbers

several inches is left between the curtain and the wall behind it.

finish materials

The materials used for finishes in the studio have a wide range of *absorptive co-efficients.*

These can be used to control the acoustical climate of the room. Co-efficients in the 125-4,000-Hz range for most commonly used materials are given in the accompanying table. Note that the absorption of almost all the materials decreases rapidly at the lower frequencies. Below 125 Hz, absorption becomes almost negligible. This reemphasizes the need for proper basstrapping in the studio.

Materials		Coefficients					
		125 Hz	250 Hz	500 Hz	1,000 Hz	2,000 Hz	4,000 Hz
WALLS	Brick, unglazed	0.03	0.03	0.03	0.04	0.05	0.07
	Brick, unglazed, painted	0.01	0.01	0.02	0.02	0.02	0.03
	Concrete block, coarse	0.36	0.44	0.31	0.29	0.39	0.25
	Concrete block, painted	0.10	0.05	0.06	0.07	0.09	0.08
	Gypsum board, ½ in. nailed to 2 × 4's 16 in. o.c.	0.29	0.10	0.05	0.04	0.07	0.09
	Marble or glazed tile	0.01	0.01	0.01	0.01	0.02	0.02
	Openings						
	Grilles, ventilating	0.15–0.50					
FLOORS	Concrete or terrazzo	0.01	0.01	0.015	0.02	0.02	0.02
	Linoleum, asphalt, rubber or cork tile on concrete	0.02	0.03	0.03	0.03	0.03	0.02
	Wood	0.15	0.11	0.10	0.07	0.06	0.07
	Wood parquet in asphalt on concrete	0.04	0.04	0.07	0.06	0.06	0.07
	Carpet, heavy, on concrete	0.02	0.06	0.14	0.37	0.60	0.65
	Same, on 40-oz hairfelt or foam rubber pad	0.08	0.24	0.57	0.69	0.71	0.73
	Same, with latex backing on 40-oz hairfelt or foam rubber pad	0.08	0.27	0.39	0.34	0.48	0.63
CEILINGS	Plaster, gypsum or lime, smooth finish on tile or brick	0.13	0.15	0.02	0.03	0.04	0.05
	Plaster, gypsum or lime, rough finish on lath	0.02	0.03	0.04	0.05	0.04	0.03
	Same, with smooth finish	0.02	0.02	0.03	0.04	0.04	0.03
	Plywood panelling, 3/8 in. thick	0.28	0.22	0.17	0.09	0.10	0.11
GLASS	Large panes of heavy plate glass	0.18	0.06	0.04	0.03	0.02	0.02
	Ordinary window glass	0.35	0.25	0.18	0.12	0.07	0.04
FABRICS	Light velour, 10 oz. per sq. yd., hung straight, in contact with wall	0.03	0.04	0.11	0.17	0.24	0.35
	Medium velour, 14 oz. per sq. yd., draped to half area	0.07	0.31	0.49	0.75	0.70	0.60
	Heavy velour, 18 oz. per sq. yd., draped to half area	0.14	0.35	0.55	0.72	0.70	0.65

**portions of this table are reprinted through the courtesy of the Acoustical and Insulating Materials Association (AIMA)*

Figure 4-20 Absorption co-efficients of common building materials

lighting

Proper lighting in the studio and control room is essential for the creation of a space that is comfortable and functional. In the studio, lighting must be powerful enough to provide musicians with the ability to see one another, communicate and read music or lyric sheets. On the other hand, it must not be so excessive that it distracts or interferes with the emotional aura of the music being recorded.

To satisfy these varying criteria, the lighting system should be *flexible*—this means that the lighting fixtures should be moveable and their brightness controlled. Very often, several circuits employing a combination of white and colored lights are used to maximize flexibility. Dimmers can be installed to control brightness.

Track systems, which allow positioning of lights as well as interchangeability of fixtures can be effective in conjunction with a fixed lighting system. Recessed lights, which create punctures in the sound seal, should be avoided in favor of surface-mounted fixtures. If a recessed fixture is used in an acoustically critical area, it can be housed in a soundproof box.

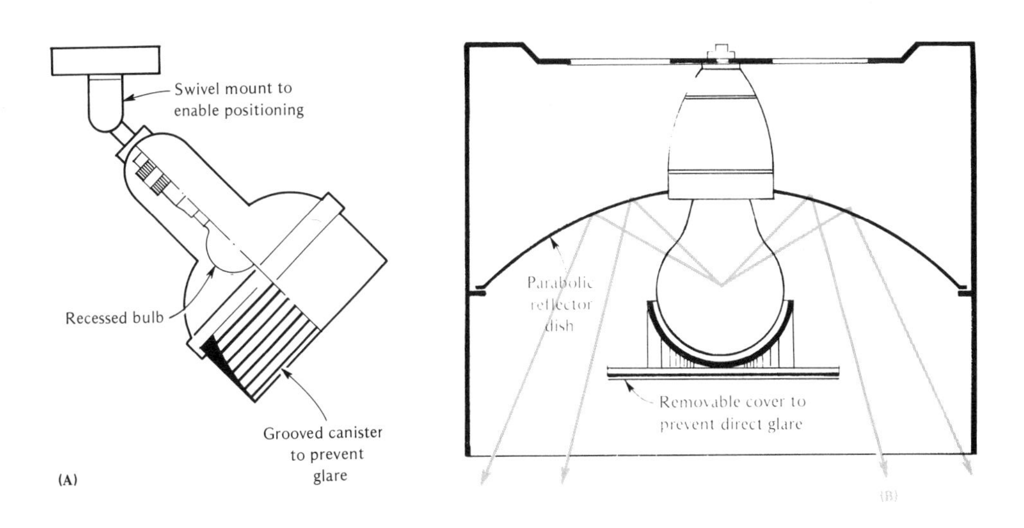

Figure 4-21 Typical lighting fixtures suitable for studio use

Common household dimmers, found in hardware stores, are normally unsuitable in studio applications due to improper shielding and transformer design—both of which can cause electrical as well as acoustical hum and interference. Typically, larger-capacity dimmers such as 1200- and 1500-watt dimmers, are better constructed and more suitable in studios, even if the lighting loads do not seem to require them. A further advantage of a larger capacity dimmer is its ability to accomodate expansion of the lighting system at a future date.

In general, fluorescent fixtures should be avoided in studios and control rooms because of their tendency to induce electrical and acoustical hum.

aesthetics

Since the look and *feel* of a room can have a tremendous effect on the behavior of the people within the room, it is no surprise that the aesthetics of a studio often have a profound influence on the caliber of music produced in it. Many studios have become famous for their "sound" which is said to be a large component of the hit records produced in that studio. Very often, this "sound" is a magical combination of acoustical phenomena and the aesthetic presence of the studio on all musicians who play inside.

Since aesthetics are subjective and enigmatic, there are no golden formulas. Most musicians agree that colors should not be too distracting. Subdued colors are often effective for setting a comfortable mood. Earthen tones such as brown, orange and tan can be warm and stimulating. However, there are countless other combinations which project inner warmth. As a rule, colors which are harmonious tend to inspire moods and music with the same quality. Colors which are flamboyant and outrageous tend to produce the opposite.

Natural materials, such as wood, rock and cork, can be especially stimulating.

Lighting, which affects our perception of color and texture, is very significant in the studio. Lights should never be harsh or produce distracting glare. Indirect and colored lights can be effective for creating moods. This seems to be especially important for vocalists and solo instrumentalists.

Natural light and views from exterior windows and skylights can also have a lovely effect—which artificial light can never duplicate. Natural light creates an open, airy feel in a studio that can be tremendously inspiring. Natural light also provides an environment in which plants and trees can flourish, giving the studio the aura of life.

chapter five

the control room

introduction

The purpose of the control room is to provide an isolated environment in which the recording engineer can operate recording equipment, balance levels, communicate visually with the musicians and accurately assess the quality of the recording. A successful control room will bring together people and equipment in such a way that the performance of each complements the other: Equipment will be visually and physically accessible. People will be accommodated in a comfortable manner that enhances both their ability to work and the acuity of their aural perception.

The acoustical properties of the control room should not color or alter the natural sound of the recording. The frequency response and reverberation characteristics should ideally be smooth and predictable. Listening conditions should never change drastically from one position to another in the room. Finally, the room must be able to accommodate a variety of musicians, producers and visitors in such a way that mutual interference is eliminated and a natural flow is maintained.

the combined studio and control room

In many home recording situations, the individual doing the recording is often the same one being recorded. In this case, more likely than not, a single room will be filled with a conglomeration of musical instruments, microphones, tape recorders and playback equipment. This room is therefore being forced to serve the dual function of a studio and a control room.

Fortunately, if the room is acoustically treated according to the methods described in the last chapter (i.e., echoes, flutter and standing waves have all been eliminated), it is well on its way as a recording room .

However, in order to maximize the accuracy of the recording room as a *mixing* environment (i.e., suitable for creating accurate assessments of the tonal balance of the recording), the following steps should be taken:

1. Designate a *mixing position* along the central axis of symmetry of the room, facing the wall on which the speakers will be mounted. For residential-sized rooms (dimensions from 10 to 25 feet), the mixing position should be separated from the wall it faces by a distance of .7 times the wall width (i.e., 7 feet away from a 10-foot wall).
2. Mount the monitor speakers rigidly to the wall in such a way that an equilateral triangle is formed between the speakers and the mixing position.
3. Mount the speakers at ear level or slightly higher, with the tweeters aimed at the mixing position.
4. Keep the monitor path clear of obstructions.
5. House all tape machines in a soundproof enclosure to eliminate motor and fan noises from leaking into the recording.

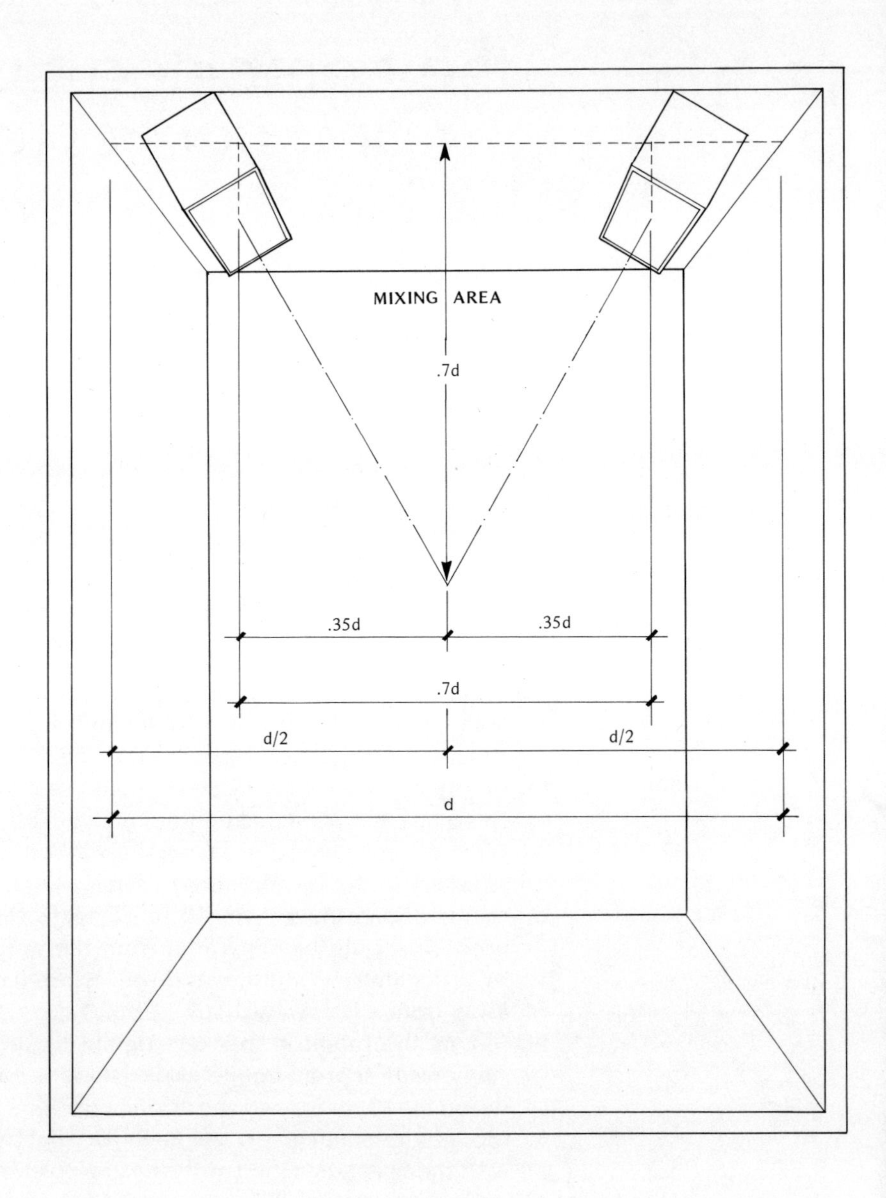

Figure 5-1 Locating the mixing position in a combined studio and control room

the optimum control room

The optimum mixing environment entails the construction of an acoustically isolated control room, adjacent to the studio. We now turn to the principle of constructing such a room.

When control rooms were first used in recording and broadcast application, they were little more than tiny square booths which accommodated an engineer and his equipment. There was little or no concern for the effect of the control room's acoustical properties on the perception of the occupants. Nor was much thought given to making the environment comfortable and beautiful.

At that time, the technical quality of the reproduction system was amateurish by today's standards. In all probability, the acoustical properties of the control room did not even have a discernible effect on the final product. The quality of most home listening equipment was too poor to detect such subtleties.

The meteoric rise of technology has changed all that. The science of sound reproduction is now so sophisticated that the best reproductions are sometimes scarcely discernible from the original performance. Also, the quality of consumer listening equipment and overall sophistication of the consumer himself (or herself) has risen dramatically. Not long ago, the consumer listened to music on monophonic equipment reproduced through a single narrow-band speaker. Nowadays, the same consumer is likely to have a sensitive stereo tuner, power amplifier and turntable, plus a set of speakers capable of relatively broadband performance. Furthermore, the rise of digital reproduction techniques is just around the corner, holding in store even higher levels of technological excellence.

As a result, the consumer is becoming acutely aware of the technical quality of the product and often will not settle for an inadequate recording. Rock stars and record companies now spend millions of dollars in search of technical perfection. Technically superlative albums are being rewarded with record sales that no one ever dreamed possible. Therefore, the role of the studio and control room in the recording process is more critical than ever before.

Designing the ultimate control room has become a rather sophisticated science in itself. Paradoxically, there is no permanent solution. As technology and music change, acoustical criteria change, and more sophisticated designs must evolve. To complicate matters, the true solution must not only satisfy technical criteria but human and psychological criteria, as well.

This chapter will not delve into all of the intricate and esoteric aspects of modern control room design, but rather present some of the basic guidelines which have been accepted and proven.

isolation

Isolation is the first task of the control room. Maintaining an average acoustic isolation of at least 50 dB (60 dB above 500 Hz and 40 dB below 100 Hz) will enable the control room to be operated relatively independently of the studio. This will allow musicians in the studio to practice and play without disrupting the control room activities. In addition, it will enable the occupants of the control room to monitor the session at loud levels without disrupting the actual recording. Acoustic isolation is accomplished via the use of decoupled walls, floating masses, air spaces and resiliently mounted windows, as discussed in Chapter III.

To maintain the integrity of the sound seal, the control room window assembly should be constructed from at least two, non-parallel panes of glass, of different thicknesses to prevent sympathetic resonance and the development of standing waves between the panes. The window should be large enough to allow complete visual access from the control room to the studio.

size

The physical size of the control room has a great deal to do with its ability to reproduce sound accurately. In order to accommodate very low frequencies, whose wavelengths can measure over 30 feet long, a proper control room must be reasonably large. For critical performance, volumes will generally be required to be in excess of 2500 cubic feet. In addition, equipment, acoustical treatments and storage will also reduce the available cubic volume considerably, thus necessitating a rather large exterior shell at the outset of construction.

The proportions of the control room dimensions also play a role in determining its performance characteristics. Since each dimension of a room has a characteristic set of acoustical resonances, various ratios of length to width to height can markedly affect how these resonances are distributed throughout the audible spectrum. For example, if a room had the same length, width and height, the standing waves forming in each of these dimensions would be identical in frequency.

These resonances would add to one another, creating a tremendous imbalance in the frequency response of the room. On the other hand, if the room dimensions were such that all standing waves forming were distributed evenly throughout the spectrum, the coloration problem would not be as severe and could be handled by the techniques discussed in the last chapter.

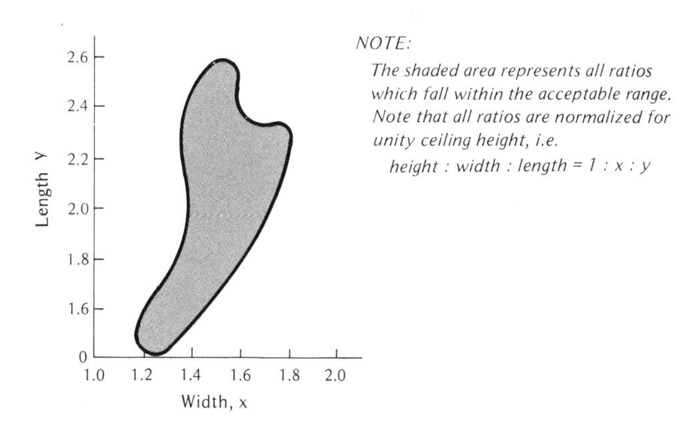

Figure 5-2 Range of acceptable room ratios (after Bolt)

Throughout history, a great many theories have been put forth as to the optimum proportions for a room in which music will be played (or listened to). In Greek times, the ratio of 1:2:3 was proposed because of its similarity to the integral multiple relationship of the fundamental and harmonic overtones of music itself. However, for exactly the same reason, this ratio turned out to cause severe standing waves that were interrelated as harmonics of one another. Other ratios have since been proposed by scientists, acousticians and musicians. It soon became evident that there was no single optimal ratio. There were, however, a range of ratios that encouraged an even distribution of resonances. These are shown in the accompanying diagram.

It should be noted, however, that by the time a control room is filled with people, equipment, furniture and acoustical treatments, the regularity of the room ratios is broken, and the ratios become less significant in determining ultimate performance than theory might predict.

geometry

To perform its complicated variety of functions, the control room must take on geometrical shape determined by acoustical as well as other criteria.

Most control rooms start with a sound-proof shell in which this geometry is constructed. The main acoustical purpose of the geometry is to provide the correct assortment of reflective and absorptive surfaces at the proper angle; to insure that sound fields are diffuse and predictable; and to eliminate standing waves, echo, flutter and room resonances.

The geometry also provides for placement of the console, tape recorders, speakers and signal-processing devices in a way that optimizes the performance of the room and its occupants.

The geometry of the room evolves from the following basic considerations: symmetry, monitor placement, equipment placement, circulation.

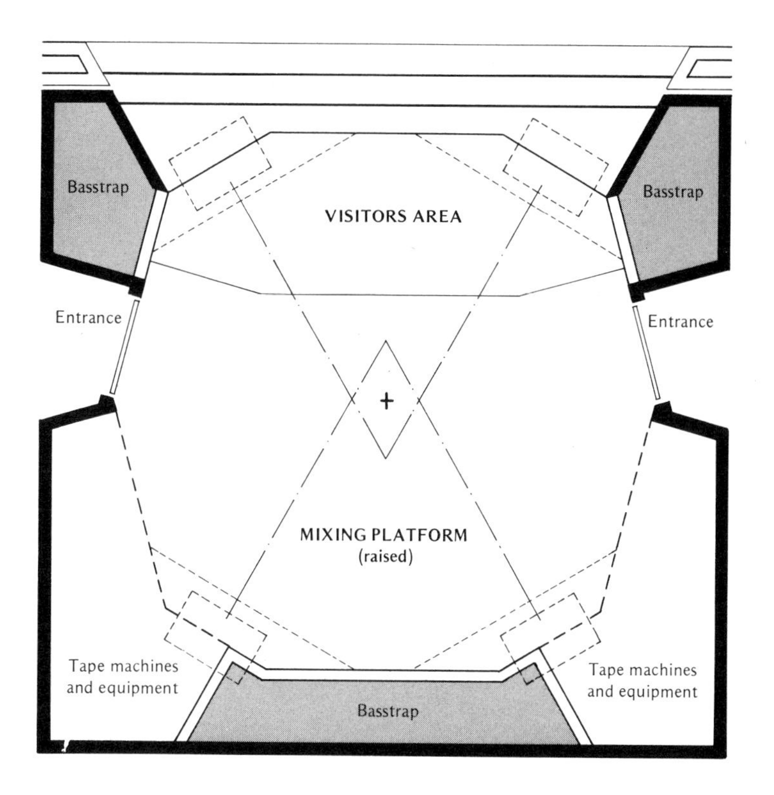

Figure 5-3 A basic control room geometry

symmetry

If sound waves from identical sources reflect differently at various points in the control room, the mixer will hear each differently. This phenomenon alters the perceived equalization, as well as the spacial imagery, and makes the achievement of an accurate stereo or quad mix next to impossible.

The problem can be avoided by constructing a perfectly *symmetrical* control room, and by positioning the mixer or engineer on the axis of symmetry.

In a stereo control room the symmetry need only be *bilateral*—i.e., along the front-to-back axis. In a quad room, the symmetry ought to be *quadrilateral* to preserve equalization and spacial imagery in all four directions.

Oftentimes, the placement of a large reflective glass pane at the front of the control room presents a problem to achieving perfect symmetry in a quad environment, as it is impractical to place a similarly blank reflective area at the rear of the control room. This problem can be partially solved by incorporating a thick curtain (which can be drawn during mixdown) into the control room window. In addition, the area of the control room window can be kept to the necessary minimum, while the area surrounding it should be kept as absorptive as possible.

positioning the monitors

Because conventional cone speakers tend to be very directional (especially high-frequency units), the placement of the monitor speakers can crucially affect the mixer's perception of the sound on tape. For example, if the mixer is positioned slightly off the directional axis of the high-frequency driver, he will not hear the full amount of high frequency actually on the program material. This will cause him to overcompensate with equalization or level—and possibly to ruin the recording.

To correct this, speakers should be placed so that their axes of directionality (usually at 90 degrees to the speakers' front) are focused *at the mixer*. This implies that the speakers should be tilted "inward" slightly (towards each other) and mounted at ear level. (Approximate ear level is 42 inches, seated; 62 inches, standing.)

If speakers cannot be placed at these approximate

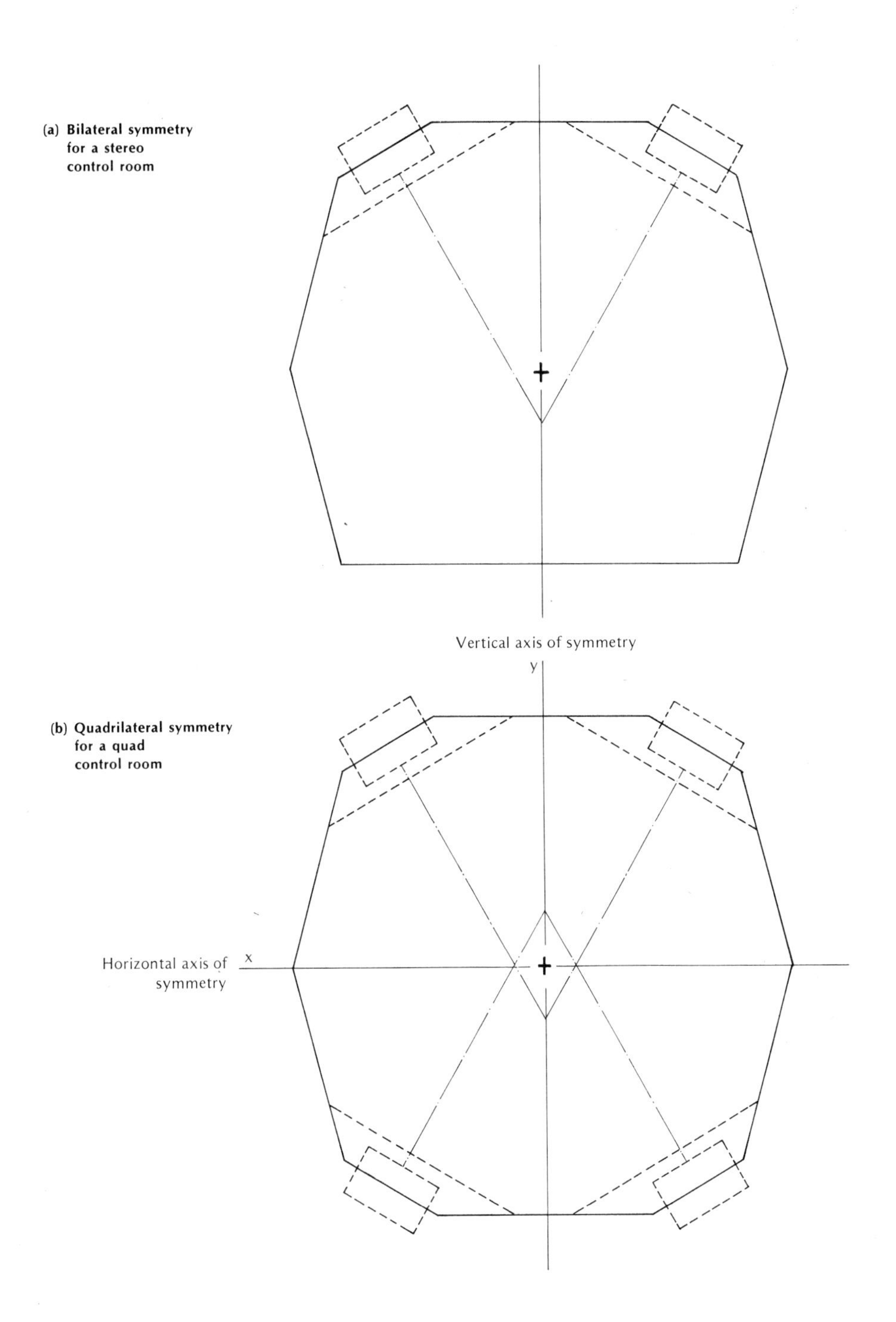

Figure 5-4(a) & (b) Symmetry of the control room

heights, accurate results may also be obtained by mounting them higher (between 6 and 8 feet from the floor) and tilting them *downward* slightly, so that their directional axis is aimed at the mixer. The axis should be drawn from the tweeter or high-frequency unit—which is the most directional of the drivers. The exact angles of monitor displacement can be determined by stretching a length of string from the tweeter position to the mixing position, and then lining the speaker cabinet at right angles to the taut string.

Actually, aiming is done to a point approximately 2 feet behind the mixer's probable location to account for the dimensions of his head and the likelihood of his horizontal movement.

the monitor spacing

If the distance between the speakers is the same as the distance to the listener, an equilateral triangle is formed. Although the exact geometry of speaker placement is a matter of personal preference, this equilateral triangle principle has been tried and endorsed by many professional engineers as well as speaker manufacturers. It is an easy geometry to install and can be accomplished very simply via a length of string or a tape measure to gauge distances.

Most engineers seem to prefer a distance of 8 to 12 feet between themselves and the monitor speaker. If the distance is less than 8 feet, they find that the speaker becomes "overbearing." At distances greater than 12 feet, it becomes too "distant." Further distances give a more reverberant sound, maximizing the contributions of the room, while closer distances give a more *direct* sound, minimizing the contributions of the room. At distances that are too close, however, the sounds of the individual monitor components do not blend properly, and the signal is heard as a conglomeration of two or three networks rather than a unified field.

To maintain true symmetry in a quad room, the rear speakers should be the same distance and geometrical relationship from the mixer as the front speakers.

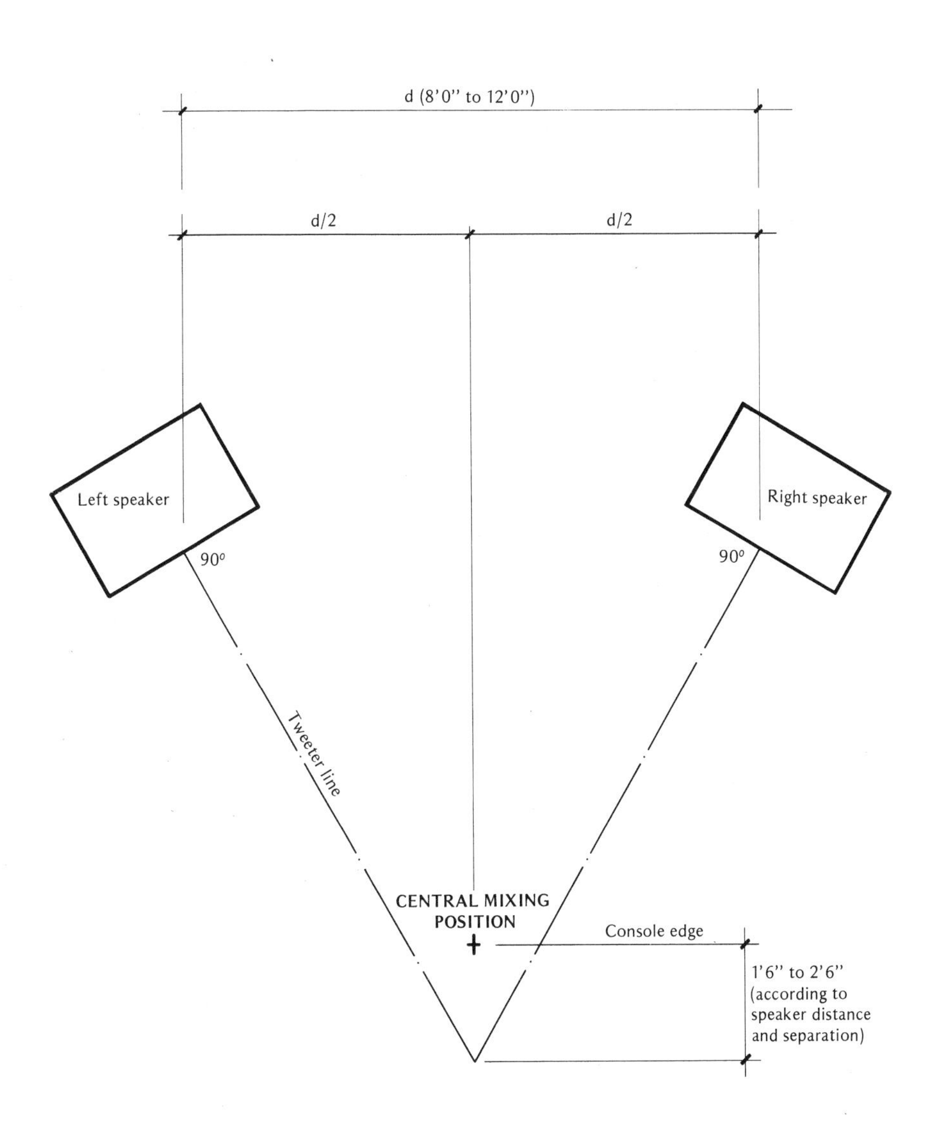

Figure 5-5 Aiming the monitor speakers

the monitor path

Any obstructions (patch bays, tape machines or mixing consoles) between the monitor speakers and the mixing engineer will adversely affect the sound of the monitor and, hence, the accuracy of the mix. A completely unobstructed path between the monitor speakers and the listening area should be maintained.

When this problem arises, the monitors can be placed *above* the control room window to a height no greater than 8 feet and tilted accordingly to achieve a direct, uninterrupted sound path to the mixing engineer.

In a quad control room, the rear speakers should be mounted at the same height as the front speakers, to maintain acoustical symmetry.

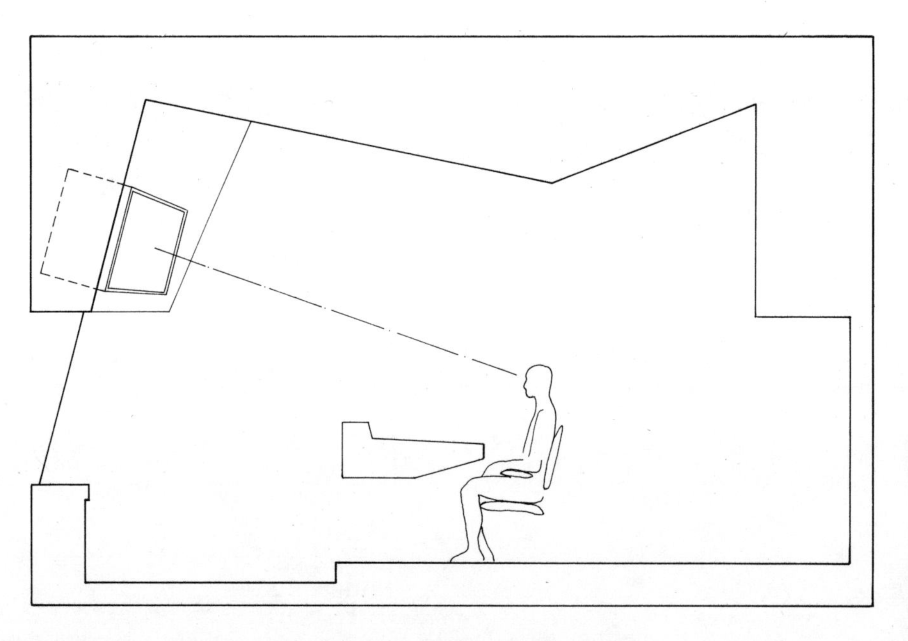

Figure 5-6 The monitor path should be unobstructed

flush mounting

The complex interaction between a monitor and its acoustical environment is very much a direct result of the physical placement of the speaker within the environment. For example, let us consider the case of the monitor mounted in mid-air, via chains or cables, versus the monitor mounted flush into the wall.

When a speaker radiates sound energy into a room, the high frequencies, which are quite directional, travel outwards in a very narrow pattern. The low frequencies, whose wavelengths can be as long as 40 feet, bend easily around the speaker cabinet, radiating to the side and rear, as well as towards the listener. Some of this energy is absorbed and thus lost. Some is reflected from nearby room surfaces and directed back towards the listener. This addition of direct and indirect sound causes phase interference in the low-frequency signal, the severity of which is a function of the distances and geometries in-

Note: Flush mounting techniques work best when used in conjunction with a well damped, massive wall such as the ***Infinite Baffle Speaker Wall™.***

Infinite Baffle Speaker Wall is a registered trademark of Jeff Cooper Architects, A.I.A.

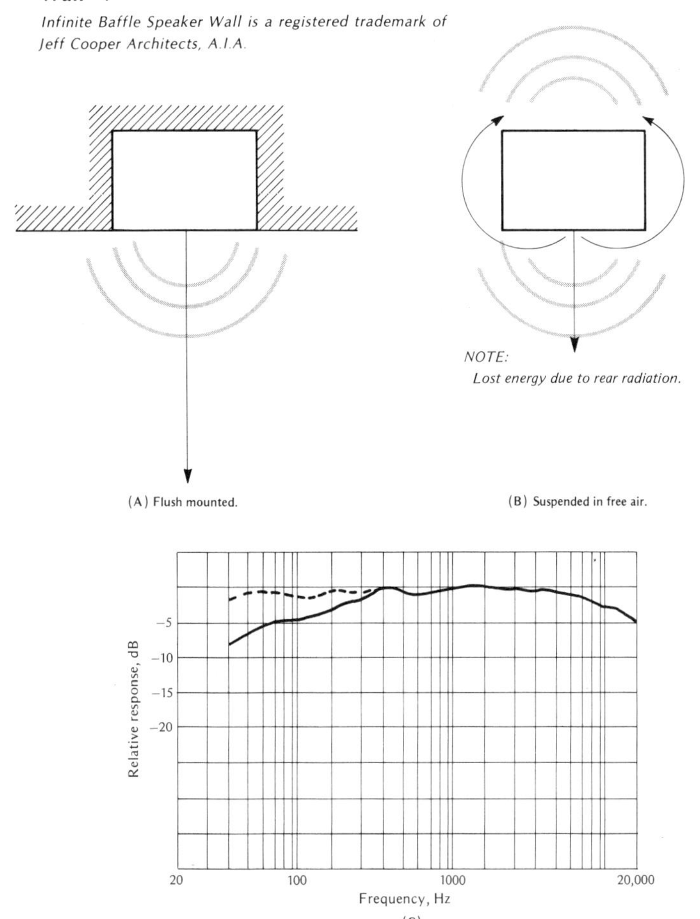

Figure 5-7 Response curve for the identical monitor speaker a) flush mounted b) suspended in free air

volved. The net effect is an unevenness and a lack of predictability of the low-frequency response of the monitor. Clearly, this is a problem.

A technique for achieving a more predictable and smoother response is to mount the monitor speaker *flush* into the wall. In this method, unpredictable phase cancellations due to reflected waves and cavity resonances may be eliminated. Secondly, a great portion of low-frequency energy, which previously was lost through *rear radiation*, may be redirected into the room, thereby improving the bass response of the monitor system by as much as 6 dB. Lastly, a flush-mounted cabinet can be well *secured* and is therefore less likely to rattle, vibrate or otherwise lose energy due to unstable mounting.

machine soffits

The surfaces of tape machines are usually metallic and thus highly reflective. To reduce the non-symmetrical reflection of sound waves off the tape recorder toward the mixer, house the machines in absorbent-lined soffits.

These absorbent alcoves also help to reduce the actual operating noise of the tape machines (scraping reels, humming motors, whirring fans, etc.). Without the provision of some absorptive surfaces nearby, noise from several multitrack machines operating simultaneously can be quite distracting.

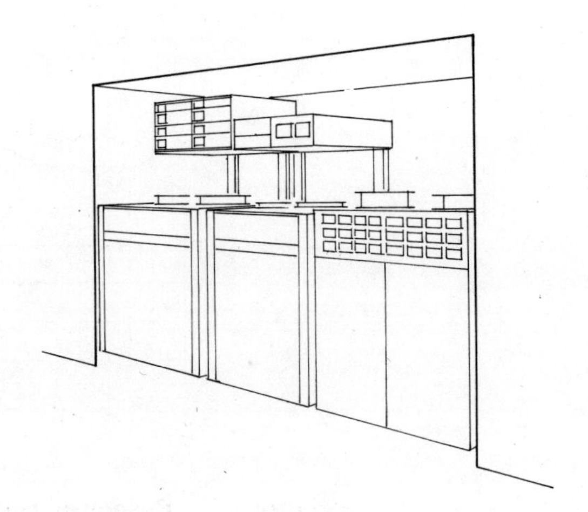

Figure 5-8 Absorbent soffits reduce machine noise

rear room reflections

When out-of-phase reflections from the rear of the control room strike the mixing engineer from behind, they can change his perception of the equalization and spacial imagery of the program content. This may result in confusion on the mixer's part as to what really is coming from the speakers.

To prevent this problem in a stereo control room, the rear of the control room can be designed to discourage the "foldback" of waves. In other words, the rear of the room would be absorptive throughout as broad a range of the spectrum as possible. This measure will also help prevent the formation of flutter echo between the reflective glass at the front of the control room and the rear wall.

This *deadening* of the rear is accomplished by installing *basstraps* and *high-frequency absorption materials* of the kind discussed earlier. The acoustic treatment should be applied to the rear and side-rear walls, if space allows.

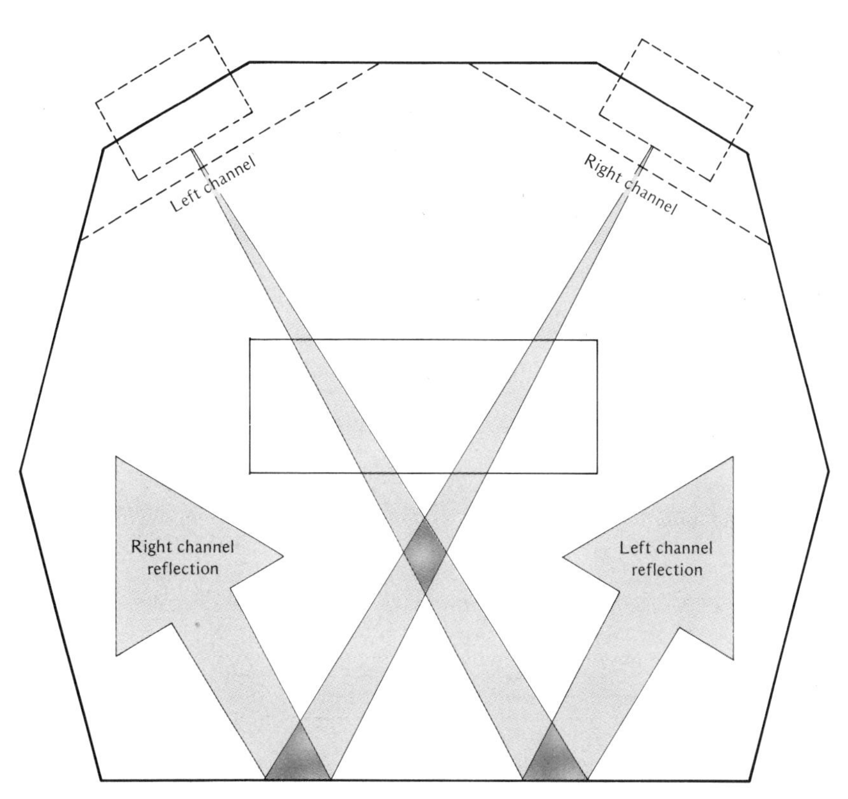

Figure 5-9 Rear reflections can cause imagery confusion

Some contend that out of phase reflections arriving at the ear within an acceptable "time window" will not adversely affect the mixer's perception. This is attributed to the so-called "Haas Effect," occasionally cited in acoustical literature. This phenomenon leads to the conclusion that reflections from the rear of the control room are not inherently problematic, when properly planned into an overall room design. Experimentation in this regard is currently under way.

control room ceiling

The geometry of the control room ceiling can dramatically affect the acoustics of the control room. In keeping with the principles discussed earlier, a control room ceiling is needed in which the geometry "breaks up" standing waves, presents a non-parallel surface to all reflective surfaces in the control room and, finally, discourages the return of reflected waves from the rear.

One type of ceiling that accomplishes these goals is the "tier-drop" or *compression ceiling* pictured below.

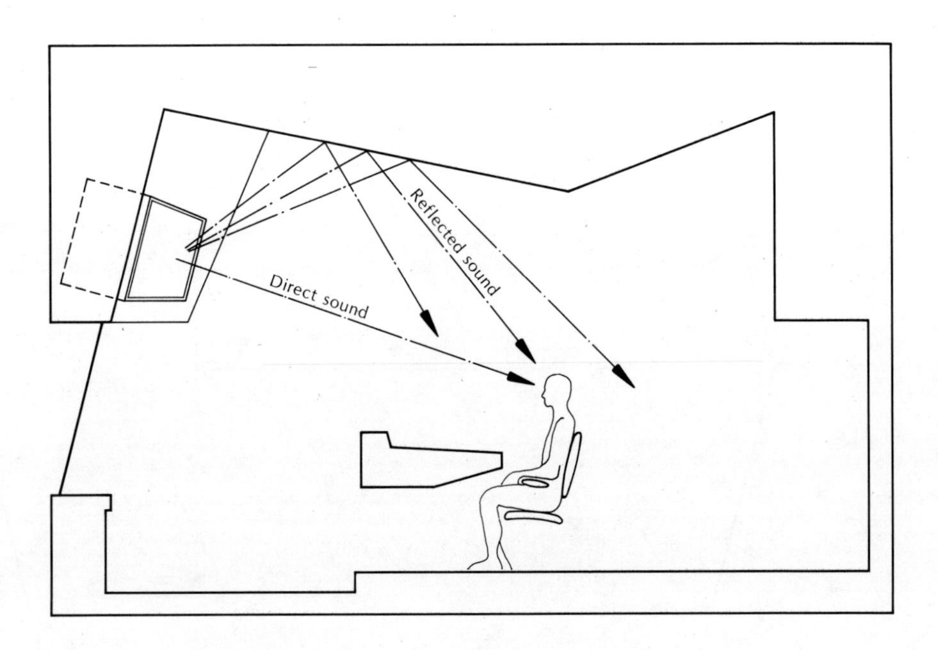

Figure 5-10 Sample control room Compression Ceiling

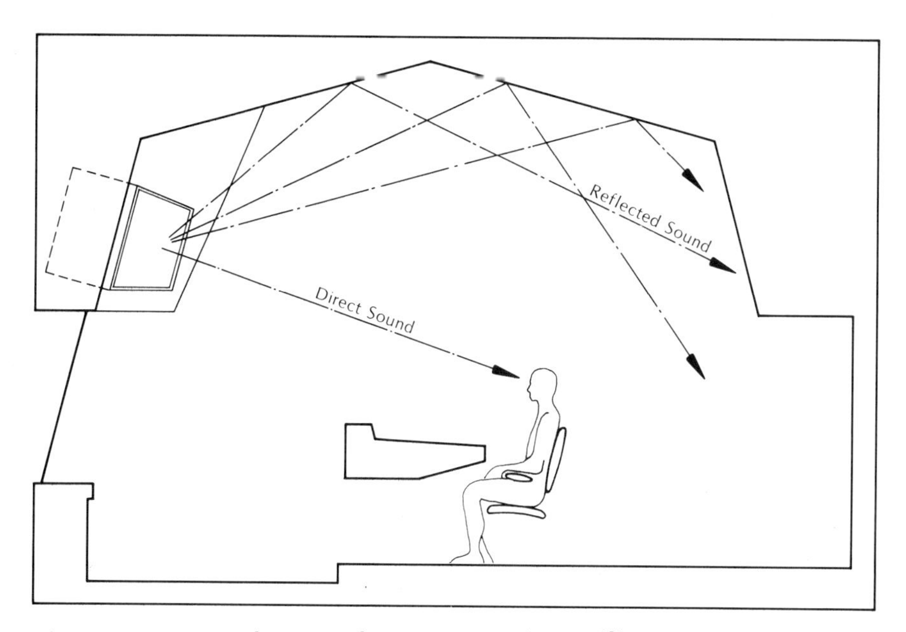

Figure 5-11 Sample control room Expansion Ceiling™
Expansion Ceiling is a registered trademark of Jeff Cooper Architects, A.I.A.

Although there are many advantages of the compression ceiling, including a general reinforcement of the bass sound, there are some acoustical effects of the ceiling which are not clearly advantageous. The first effect results from the way early reflections from the ceiling combine with the direct sound of the monitor at the mixing location. The path length difference between the direct and reflected waves causes phase cancellation to occur at a multitude of frequencies, as well as a phase reinforcement to occur at a series of other frequencies and their harmonics. The net acoustical effect is similar to that of a *comb filter* applied to the monitoring system. To complicate matters, the active frequencies of the filter constantly change with the position of the mixer. This phenomenon can result in perceived flanging and phasing effects, which can be detected by a trained ear.

A new ceiling has recently been developed to solve these problems while retaining some of the advantageous features of the tier-drop ceiling. Known as the *Expansion Ceiling*™, this type of construction results in less phase interference and excellent stereo imagery. This ceiling is illustrated above in Figure 5-11.

The *front sloping portion* of the Expansion Ceiling™ is normally designed to be as rigid as possible, using several layers of plywood, gypsum board and other materials, in order to extend the bass reinforcement effects of the front speaker baffle. The outer visible surface of the front portion is usually treated with absorptive material, such as compressed fiberglas with fabric covering, to discourage early reflections and the resulting mid and high frequency phase cancellation of the type discussed previously.

The *rear portion* of the Expansion Ceiling™ is usually vented into a large broadband bass trap in order to keep reverberation time of the control room in the low frequencies to a minimum and to produce a well defined or *tight* bass sound from the monitor speakers.

The rear portion can also be used to selectively *tune* the control room by application of a thin membrane to the outer surface, in order to eliminate specific standing waves characteristic of a particular room. Generally, membranes can be designed to absorb very selective low frequencies when used in conjunction with a large airspace. The deep multidimensional airspace of the Expansion Ceiling™ combined with its large overall surface area gives us great flexibility in this regard.

One advantage of this type of *architectural tuning* procedure is that significant final adjustments can be made to a room's *response curve* after the basic construction is complete, for relatively little cost, by experimenting with the membrane in order to obtain the flattest frequency response.

other functions: visitors

In addition to providing an accurate listening environment and housing equipment, the control room has to accommodate musicians and visitors to the session. This is usually accomplished by allocating a specific area in which visitors sit, relax and listen to the music. It is important that this *visitor's* area is constructed in such a way that it does not interfere with the engineer's vision to the studio or his access to recording equipment. Therefore, the visitors' area is often located directly in front of the console, sunk 6 inches or more to enable the engineer to have an unimpeded view of the studio. It is helpful to treat the visitor's area with absorptive acoustical materials to minimize the disturbance of conversations and other noises of the visitors.

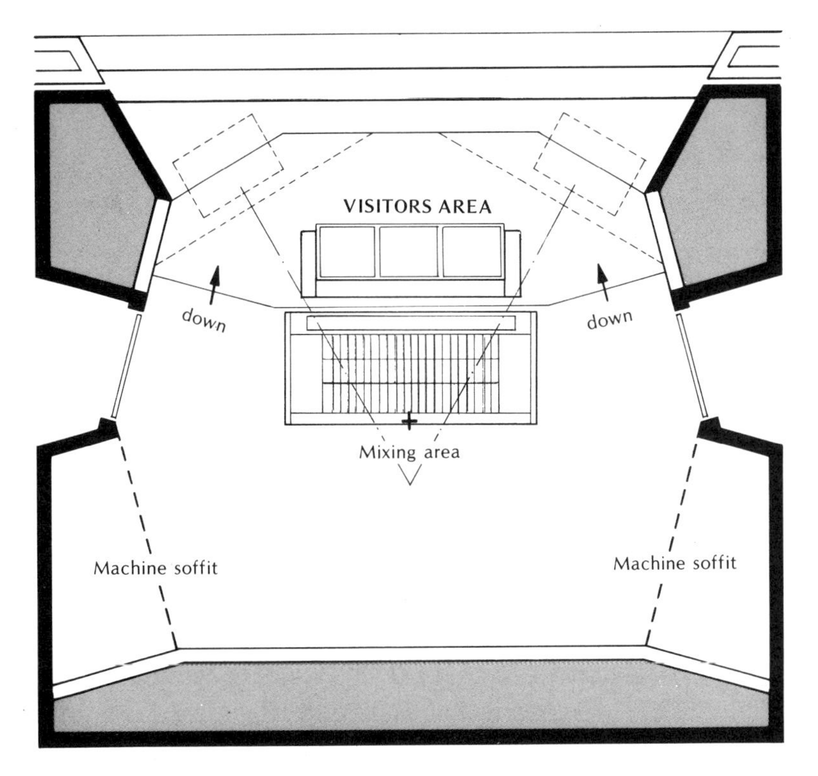

Figure 5-12 Visitors area

interface

In order for the various electrical components of the control room to function as a system, they must be interconnected, often in multiple ways. Mike lines must be routed to the console, which in turn is connected to the tape machines, monitoring equipment and other devices. The process of wiring the studio is often referred to as the *interface*.

Many of the devices used in the studio and control room will require some type of power to operate. In the studio, the instrument amplifiers and lights will require connection to the building power main. In the control room, the tape recorders, amplifiers and other auxiliary devices will also derive their power from this power main, as will the lights. Even the console, which is powered by an internal power supply, will ultimately be connected to the power main in the building.

Consequently, there will be many relatively high-voltage power lines crisscrossing the control room and studio. At the same time, there will be a great deal of audio wiring interconnecting the console, tape machines and other devices. The power lines will be carrying *high-voltage signals*, whereas the audio lines will be carrying extremely *low-voltage signals*. If care is not taken in the placement of these wires, it is quite possible that the high-level signals will interfere with the low-level signals.

The most notable interference effect would be the induction of a *60-cycle hum* in the audio signal. This effect could be disastrous in the recording.

To eliminate 60-cycle hum, or to minimize the probability of its occurrence, proper grounding and shielding are essential. All connections should be well-soldered, and the grounding wires should be connected to a common grounding plate, which is ultimately wired to a buried copper plate or copper water pipe.

troughs

Another important precaution is to physically separate wires carrying high-level signals from wires carrying low-level signals. This is accomplished in the studio and control room by providing separate wiring chases or *troughs* for each group of wires. These troughs should be a minimum of 12 inches apart.

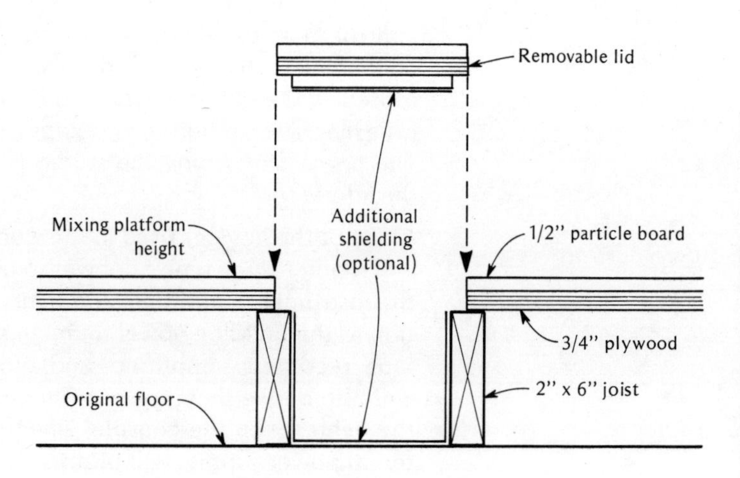

Figure 5-13 Typical trough construction

Occasionally, it will be necessary to cross the paths of high- and low-level signals, especially if the wires are being concealed in a common floor. In this case, if the wires are crossed at an angle of 90 degrees, interference will be minimized owing to the greatly reduced intensity of the electromagnetic field in this orientation.

radio frequency interference

Occasionally, low-level signals may also be affected by the presence of airborne radio signals. This type of interference is normally called *RF (radio frequency)*. Simply stated, the low-level signal wire acts as an antenna, capturing the radio signal. The many amplifier stages through which this signal passes amplify the interference, contributing to its audibility. Proper shielding and grounding are normally a solution to this problem.

In severe cases, such as a studio located in the direct path of a radio transmitting tower, the interference can only be eliminated by surrounding the control room in a wire shield such as copper mesh, concealed in the walls, ceiling and floor. This mesh would, in turn, be grounded to a copper plate buried in the ground.

tuning the room

Since the performance of any monitoring system is affected by so many exterior factors (speaker placement, room geometry, air temperature, humidity, etc.), it is normal for even the best of studio monitors to have small irregularities in performance when placed in the control room. In an effort to smooth out these irregularities and make the monitor more compatible with the environment in which it is placed, the monitor can be "tuned" to the room.

Tuning amounts to *compensating* for the irregularities in the combined monitor/room output by using controlled amounts of electronic equalization. Therefore, if the performance of the monitor is exaggerated at one frequency (i.e., its output is significantly *higher* than other surrounding frequencies), the correct amount of equalization is installed in the circuit and the output of the monitor at this frequency is reduced. Similarly, if the output at a certain frequency is *lower* than surrounding frequencies, the correct amount of EQ is installed and the level is raised. The result is a monitor/room frequency-response curve that is more flat (less irregular) and thus more accurate.

In practice, equalizers centered at *third-octave bands* are usually used. This allows adjustments to be made in very narrow areas throughout the audible spectrum. This system of tuning is called *"third-octave-band equalization"* and is used extensively in professional recording studios.

To determine the amount of equalization necessary, tests are conducted with a sound-level meter and a third-octave-band filter set—or a microphone and a real-time spectrum analyzer that exhibits the monitor response visually on a screen or through an L.E.D. display. Because of its random nature, *pink noise* is normally used as a source, although other techniques do exist.

Measurements are taken in *several* room locations to give a more representative picture of room response, although the most critical measurements are taken in the immediate vicinity of the mixing console at the engineer's position.

A room need not necessarily be tuned "flat." Occasionally, the response of a control room will be gently "rolled off" (i.e., the output is gradually reduced) at very high- and low-frequency bands, so that the output of the

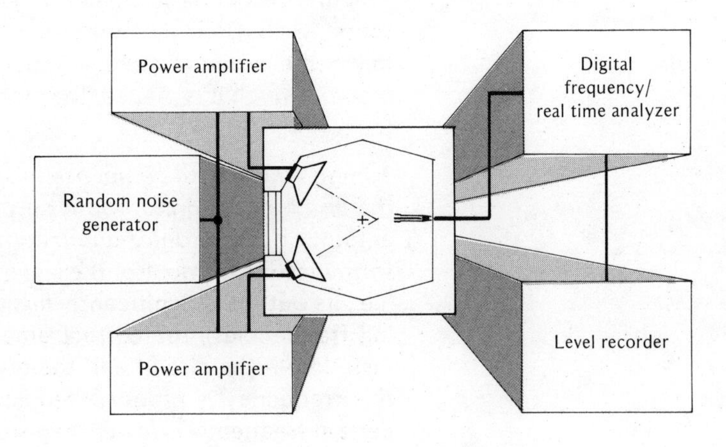

Figure 5-14 A typical control room test set-up

monitor will more closely approximate typical consumer listening equipment, which generally lacks *extreme* highs and lows. In this way, mixes can be optimized for the consumer.

Some engineers believe it is better to actually bring a pair of consumer-type loudspeakers into the control room to test compatibility of the mix for the consumer medium. In this way, the control room monitor itself can be optimized for supreme accuracy.

The problems with third-octave-band equalization

Although third-octave-band equalization can certainly help smooth out the performance of the monitor, it is not a cure-all for the acoustical problems of the control room. In essence, electronic equalization can never truly compensate for a problem that is inherently *acoustical.*

For example, the *exaggerated* response of a monitor at a particular frequency is often the result of a lengthened reverberation time at that frequency in the room. Since

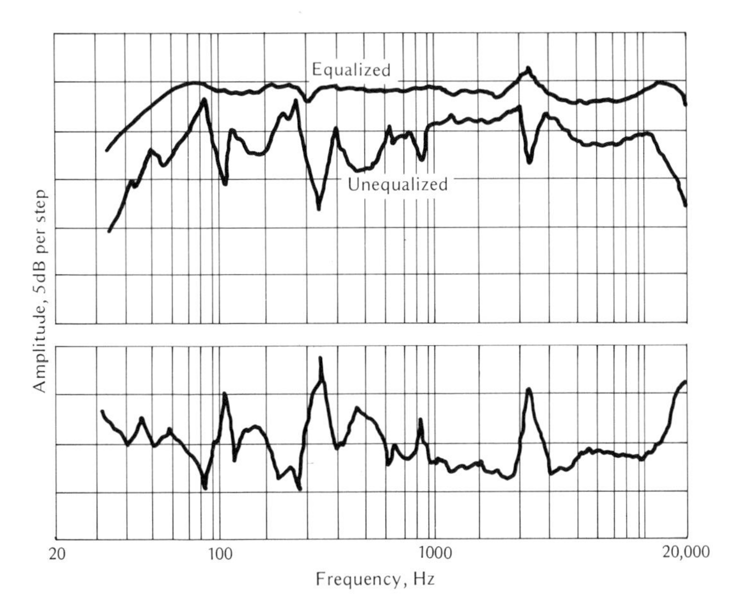

gure 5-15 Actual results of room testing and ualization

Speaker response in an anechoic chamber.

Same speaker response in an untreated control room.

Same room and speakers after the application of selective absorption (no electronic equalization).

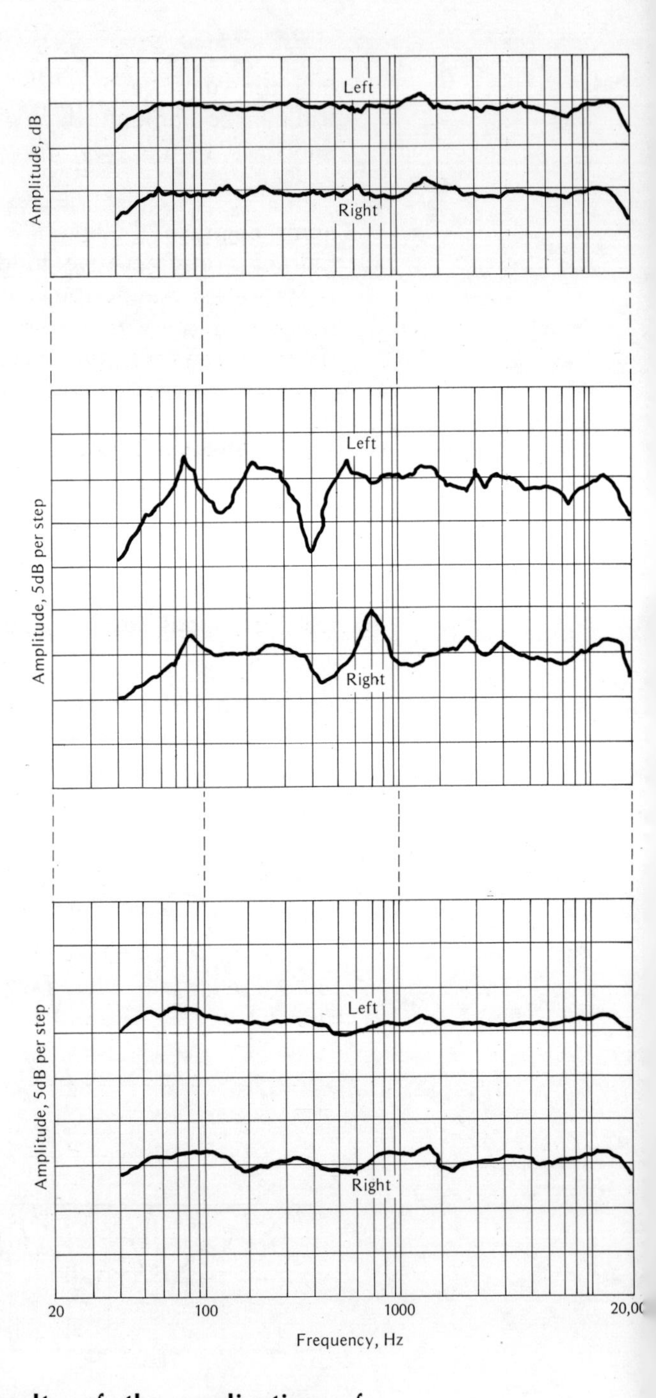

Figure 5-16 Actual results of the application of selective absorption in the control room for the purpose of fine-tuning the room

the microphone (which tests the monitor output indiscriminately) adds the reverberant sound of the room with the direct sound from the monitor, the test results will show a peak as a result of the extended reverberation. In actuality, the monitor itself may be perfectly flat at that frequency. Going by the test results, however, the output is determined to be high and a compensating amount of equalization is installed. *Unfortunately, this does absolutely nothing to eliminate the basic reverberation problem.* Prolonged reverberation at that frequency will still cause sounds to linger and result in coloration. Furthermore, the output of a monitor that may have been flat in the first place now has been altered with an intentional irregularity.

As a second example of the problems with third-octave-band EQ, consider the following: *Reduced* output of the monitor is often the result of phase cancellation occurring between the direct and reflective sound waves in the control room. The installation of EQ may make the monitor *look* flat on an analyzer, but does nothing to affect the phase cancellation.

Lastly, many feel that the insertion of equalization into the circuit of a monitor system (especially in large amounts) often causes a certain unnatural sound associated with harmonic distortion and "ringing," among other phenomena.

These reasons all suggest that equalization is not the ultimate answer. A much more intelligent approach is to correct *acoustical* problems in the room with *acoustical* solutions, rather than resorting to electronic compensation. This implies installing absorptive and reflective surfaces whose position, surface area and frequency response have been specifically designed—an exact science that demands extensive testing and mathematical calculations. As such, it is beyond the scope of this book. If an absolutely critical and natural control room monitor system is desired, an expert in this area should be consulted.

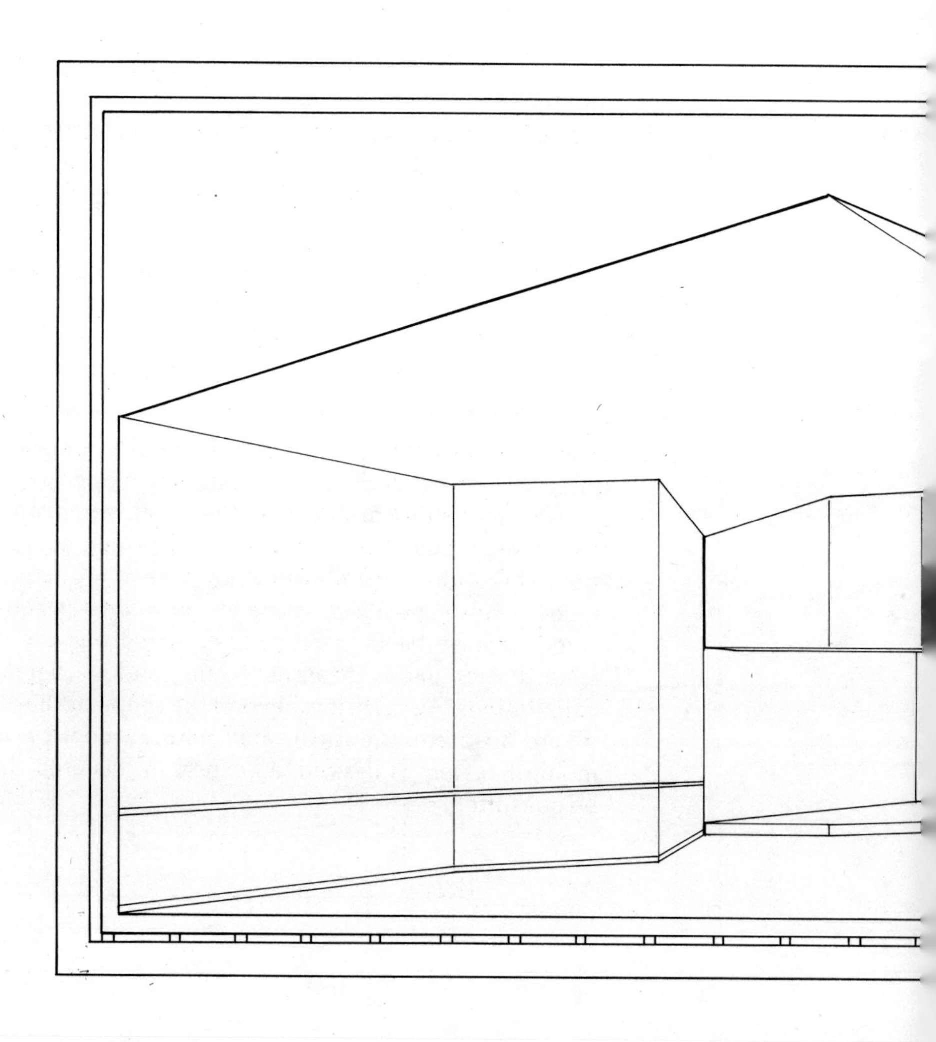

evaluating the room

After all final adjustments have been made, the control room will ultimately be evaluated by how it *sounds*. Subjectively, an excellent control room should be transparent or acoustically colorless. It should be capable of reproducing large masses of polyphony, while simultaneously retaining exact definition of individual sounds. Its *bass* should be powerful, solid and thudding—felt as well as heard—while its high frequencies should be delicate, shimmering and crisp.

The *components* of the monitor (woofer, midrange, tweeter) should not be differentiated from one another; transition at the crossover frequency should be smooth, not detectable. The sum of all components should be a unified, cohesive sound field, not a series of competing ones.

Stereo panning of the signal should result in a source whose position can be easily pinpointed. Horizontal stereo imagery should be exact. *Smearing* (broadening of the perceived source) should never occur when the signal is panned to the center. Lastly, high-frequency response should be smooth over a horizontal panorama of at least 60 degrees, while simultaneously maintaining at least 30 degrees of vertical dispersion to allow accurate listening throughout the room environment.

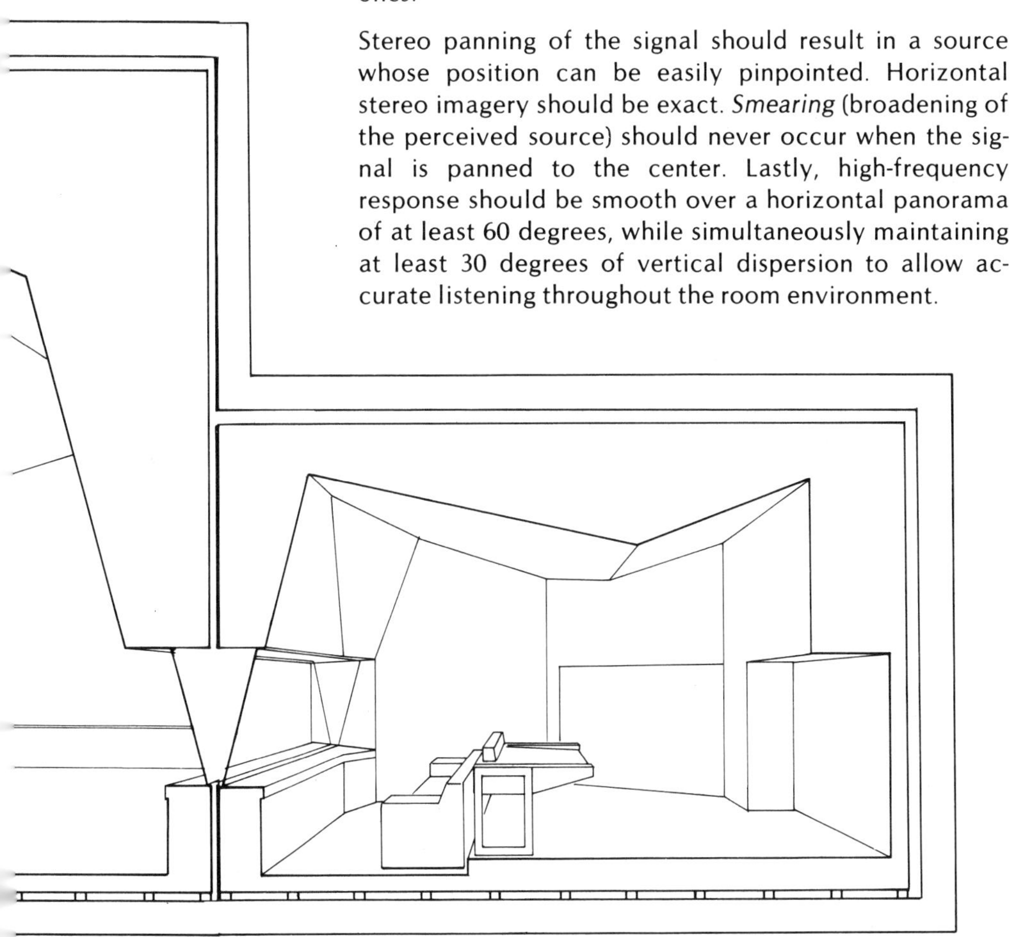

glossary

absorption The dissipation of sound energy into another form of energy—usually heat. When a sound wave encounters resistance, such as that provided by a wall or other obstacle, absorption occurs. Absorption is measured in *sabins* (after Wallace Clement Sabine). One sabin is the amount of absorption offered by one square foot of open air.

absorption co-efficient The portion of energy absorbed when a sound wave strikes a material. The absorption coefficient of a material is dependent on the frequency of the sound waves and can vary from .8 or .9 (for acoustical tile) down to .01 (for highly polished marble) at 1,000 Hz. A table of absorption co-efficients is given on page 142.

absorptivity The same as absorption co-efficient.

acoustical tile A porous architectural material, usually constructed from fiberglas or pressed particle board which is highly absorptive at the high frequencies (over 1,000 Hz).

acoustics The study of sound waves and how they behave in rooms.

ambience The residual "room sound" of a musical environment. In classical music, for example, the term ambience is used interchangeably with the word reverberation, to refer to the persistence of decaying sounds in the concert hall. (Not to be confused with *ambient noise,* which refers to background noise.)

ambient noise The average amount of background noise in an environment, measured in dB.

amplitude

The height of a waveform above or below the zero line. In this book, the term is used predominantly to describe acoustical phenomena. In this context it refers to fluctuations of air pressure above and below normal atmospheric pressure. This corresponds roughly to loudness. However, the term amplitude can also be used to refer to electrical phenomena, in which case it represents the fluctuations of voltage or current.

anechoic

An extremely dead acoustical condition in which reverberation is non-existent (i.e. anechoic = no echo). Because of the absence of acoustical character in such an environment, specially designed rooms called *anechoic chambers* have been developed in which critical audio testing can be done (see *anechoic chamber*).

anechoic chamber

A specially designed room in which all sound reflections are absorbed and reverberation can not exist. Such spaces are utilized extensively for the testing of microphones, loudspeakers and other audio equipment.

attenuation

The reduction of amplitude and therefore sound level. The term can be used acoustically and electrically. In rooms, attenuation is provided by acoustical barriers, such as walls or baffles. In mixing consoles, for example, attenuation is provided by volume controls or faders.

automation

The computerization and automatic control of the normal mixing functions (i.e. volume level, effects, equalization). Automation is made possible by storing information about the mix (level settings, etc.) on a separate track of the multi-track tape. This track is subsequently used to retrieve and operate the mixing functions independently, via the use of voltage controlled amplifiers (V.C.A.'s).

baffle

A device used to inhibit the propagation of sound waves. Baffles or "gobos" are usually employed to prevent *microphone leakage* between two instruments being recorded simultaneously. They can be constructed from wood, fiber board, gypsum, or lead, used in conjunction with sound absorbing materials, and should be made portable, for increased flexibility.

bandwidth

The frequency spectrum between two given reference frequencies. For example, the octave bandwidth between 125Hz and 250Hz would refer to *all* frequencies between 125 and 250Hz. Audio component specifications are often given with reference to a specific bandwidth (i.e. 2dB from 15 to 15,000 Hz).

basstrap

A low frequency absorber. Because of their large wavelengths (up to 50 ft.), low frequencies are particularly difficult to absorb. Therefore, special constructions must be designed in both the studio and control room to "trap" these waves, if unwanted leakage and room resonances are to be eliminated. These constructions are called basstraps. There are many techniques for designing and building basstraps, some of which involve extensive mathematical calculation and rather intricate construction techniques. The simplest type of basstrap, and therefore one of the most commonly used, is the *membrane absorber* which utilizes the dissipating motion of a vibrating membrane, such as plywood, to absorb unwanted sound.

cancellation

The destructive interference of two or more sound waves. Waves of similar frequencies and amplitude, but of opposite phase (180°) can produce extreme mutual cancellation effects.

close-miking

A technique where the mike is placed very close to the instrument being recorded to eliminate the pickup of unwanted background noise or other instruments.

compression

The reduction of dynamic range. Compression is accomplished by means of a compressor or limiter.

constructive interference

The addition of two waveforms of similar phase. Constructive interference is responsible for the production of standing waves in which a signal and its successive reflections are continually added to one another.

critical distance

The distance from a sound source at which the *direct energy* (energy radiated directly from the source) is equal to the *reverberant energy* (radiated from walls, floor and ceiling) (see *direct field, reverberant field*).

cut-off

The frequency at which a given effect ceases to operate. For example, a high-pass filter might have a low end cut-off frequency of 1000 Hz, so that all signals below 1000 Hz would be attenuated. Practically speaking, exact cut-offs are rarely desirable, for they would cause noticeably distinct and unnatural alterations of the sound. Instead, cut-offs are usually graduated or "rolled-off" in steps of several dB per octave (3, 6 or 12 for example) beginning *at* the cut-off frequency. In this way a smoother and more natural sounding transition is made.

cycle

A complete to and fro movement of a vibrating sound source corresponding to a high and low pressure wave.

dead

An acoustical condition in which reverberation is absent. A room whose surfaces are covered with heavily absorptive materials, such as thick carpeting and curtains and ample basstrapping would be referred to as dead.

decay time

The length of time taken for a signal to decay to a specific portion of its initial value. Decay time is often frequency dependent. For example, the decay time of a room at a specific frequency is the time necessary for a sound of that frequency to decay to one millionth of its original level (i.e. 60 db of decay).

decibel (dB)

The measuring unit of sound pressure and hence loudness. The decibel is actually a numerical ratio between the sound pressure of a given sound and the sound pressure of a reference sound (usually .0002 microbar). Common decibel levels encountered vary from the rustling of grass (15 dB) to conversation (50 dB), to live rock groups (110 dB) to jet plane engines at close range (130 dB).

density

The ratio of weight to volume, measured in pounds per cubic foot, or grams per cubic centimeter.

destructive interference

The addition of two waveforms of *different phase* resulting in an output of more negative amplitude than either of the original waveforms. Two identical waveforms, slightly out of phase, can interfere both constructively and destructively, at different points.

diaphragm

The delicate membrane used in dynamic microphones to sense incoming sound waves. In the microphone the diaphragm is attached to a coil in a magnetic field, which moves with the diaphragm when affected by a sound pressure wave. The movements of the coil induce a current, which becomes the mike's output.

diffraction

The bending of a sound wave around an obstacle. Low frequency sound waves, which can measure up to 50 feet long, diffract easily in the studio, contributing to the phenomenon of leakage.

direct field

The area in a room which is in the immediate vicinity of the sound source. Because of the close proximity of the source, the acoustical characteristics of the room have little influence on the sound quality in the direct field (see *reverberant field*).

direct sound

Sound waves arriving at the listening location directly from the source. To be distinguished from *reflected sound,* which arrives at the listening location after bouncing off the surrounding surfaces.

discrimination (front to back) The measurement used to describe the relative sensitivity of a mike to sounds from the rear. It is defined as the attenuation of a signal *180° off-axis* with respect to an equally loud, equally spaced signal (0° *on-axis*).

distortion (harmonic) The production of harmonics which do not exist in the original waveform.

drum booth An enclosure designed to help reduce recording leakage from the relatively high sound pressure levels of the drums; while still allowing the drummer to maintain visual and musical contact with the rest of the musicians. A properly designed drum booth is an intricate construction utilizing floating platforms and overhead canopies, which can be used to "tune" the drum sound to a desired tonal quality.

drywall An architectural wall construction not utilizing plaster or other "wet" application techniques. Drywall materials are usually manufactured in 4′×8′ panels and are applied to wood or metal studs by the use of nails or similar techniques. There is a great variety of drywall products commercially available carrying names such as *gypsum board, plaster board* and *gypsum drywall*. Although these materials vary slightly in composition (depending on manufacturer), most are similar with respect to their soundproofing ability; providing their density and thickness are equivalent.

echo A distinctly discernible reflection or repetition of the source signal. The term is often used incorrectly to refer to *reverberation*, which consists of *multiple* decaying reflections so closely spaced that they are indistinguishable from one another.

equalization The adjustment of timbre, or tone quality, achieved by changing the amplitude of a signal at different frequencies. (Abbreviated: EQ.) The tone controls on a hi-fi set are simple forms of equalization. More complex equalizers incorporate adjustable frequency centers and bandwidths, for increased flexibility.

equalization (third octave band) Equalizers divided into the third octave bands centered at 16, 31.5, 63, 125, 250, 500, 1,000, 2,000, 4,000, 8,000, and 16,000 Hz. This grouping allows flexible equalization over the entire audible bandwidth.

equal loudness contours A set of curves of equivalent loudness, which model the ear's frequency response throughout the audible spectrum. The curves, obtained from actual testing, show how much more sound power is required at one frequency than another to obtain a sound of equal loudness. The results show that the human ear is less sensitive to sound at the extreme high and low frequencies.

fader A linearly operated potentiometer or volume control. Faders are used almost exclusively for volume adjustments in mixing consoles, instead of rotary potentiometers, because of their ease of operation and visual correspondence to the level setting.

Fletcher-Munson curves The Equal Loudness contours plotted by the researchers Fletcher and Munson.

frequency The speed of vibration of a sound wave, measured in cycles per second or hertz. Frequency determines pitch; the faster the frequency, the higher the pitch. The human ear can hear frequencies in the range of 20 to 20,000 Hz.

frequency center The center of the bandwidth of frequencies.

graphic equalizers

Equalizers which provide a visible display of the frequency response curve by the relative position of the knobs on the unit. Graphic equalizers usually provide control over many frequencies simultaneously, with each knob being able to provide cut or boost of up to 15 dB per frequency.

harmonic distortion

The production of harmonics which do not exist in the original waveform.

harmonics

The overtones of a fundamental note. The first harmonic of a note is the note itself. The second harmonic is the note doubled. The third harmonic is the note tripled . . . and so on. (i.e. 500, 1,000 Hz).

hertz

The measuring unit of frequency or the speed of vibration of a sound wave. Synonymous with "cycles per second" (C.P.S.).

homo-geneous

Uniform in structure and composition.

npact isola-n class (IIC)

A system for rating the ability of a structure to isolate *impact noise* (i.e. footsteps, and other *vibrational* disturbances). Normally used in reference to floor and ceiling constructions, the IIC method utilizes whole positive numbers for rating purposes.

npact noise

The noise heard as a result of vibrations transferred through the structure of a room. Foot thumps are impact noise.

npact noise rating (INR)

A system for rating the ability of various constructions to isolate impact noise. The INR method utilizes a single number rating system, to facilitate the comparison of several structures. The INR method of rating impact noise is being replaced by the IIC (Impact Isolation Class) method.

incident (energy) The sound energy arriving at or striking a surface.

intensity The amount of sound energy radiated per unit area, measured in watts per square cm.

live A reverberant acoustical condition, usually used in reference to a room whose many reflective surfaces encourage a lengthy reverberation time.

logarithm A mathematical operation which compresses a large range of numbers to a smaller range. In the base 10, the logarithm of a number is simply the power to which 10 must be raised to produce that number. For instance, the log of 100 is 2 since $100 = 10^2$. The log of 1,500 is 3.2 since $10^{3.2} = 1{,}500$. What is the log of 10,000? (ans. 4)

mass law The law of physics which states that a material's ability to reduce the transmission of sound is proportional to its weight. According to the mass law, to decrease a wall's transmission by 6 dB it is necessary to double the thickness of the wall.

microbar (abbreviation: bar) The basic measuring unit for atmospheric pressure and therefore sound pressure. Microbar is a measure of force per unit area (i.e. 1 microbar = 1 dyne/cm^2). Since sound consists of a rapid *fluctuation* in air pressure, the measurement of a "sound pressure" actually refers to its *RMS* value. At 1 kHz the approximate threshold of hearing is 2×10^{-4} microbar (i.e. .0002 *u* bar). This value is used as a reference level by which all other sounds are measured. (See *decibel for further explanation.)*

microphone

A device which converts sound pressure waves to electrical impulses. Because it converts energy from a mechanical form to an electrical one, a microphone is called a mechanical-electrical transducer. The three basic types of microphones used in recording are the dynamic, ribbon and condenser.

mixdown

The process of balancing and combining the tracks of a multi-track recording. Mixdown is accomplished by connecting the outputs of the multi-track recorder (i.e. 4, 8, 16 or 24 track) to the inputs of a mixing console which, in turn, feeds the master machine (i.e. 4 track for quad, 2 track for stereo, full track for mono).

mixing console

The console used to balance and combine the sounds of several sources in a recording. During a *session*, the console can be used to combine the outputs of several *microphones* or instruments onto a single track of the multi-track recorder. During *mixdown* the console can be used to combine the tracks of the multi-track recording onto a master tape. In addition to volume adjustments, most professional mixing consoles allow the flexibility of *altering* the signal by adding selected amounts of equalization and reverberation as well as other effects. Most consoles also provide a talkback system (for communication with the studio) and a separate monitor section for allowing the engineer to monitor levels independently of the mix.

monitor

The control room reference loudspeaker. Because the monitor system is used as the standard by which all sounds in the control room are judged, it is important that the monitor contain no arbitrary "peaks" or "dips" in response and that the monitor-room frequency response curve be "flat," especially in the listening area.

Note: The word "monitoring" is often used synonymously with the word "listening" in studio terminology, e.g. "While monitoring, we noticed some distortion in the guitar track."

mono Common abbreviation for "monaural," meaning from a single source. Since most pop music receives its initial exposure on AM radio, which broadcasts monaurally, it is important that mixes be made in mono as well as in stereo to insure the integrity of the mix. When a stereo tape is played monaurally the center images will appear to boost by 3 dB.

multitrack The capability of recording as many as 40 individual synchronized tracks (each representing an original musical performance) on a single piece of tape. As a result of multitrack technology, many instruments, though played at different times and perhaps in different studios, can be combined to produce a single recording which gives the impression that all instruments and voices performed simultaneously. Most pop and rock music recording utilizes multitrack technology to expand its creative limits (see *overdubbing*).

noise Unwanted sound.

notch filter A filter of extremely narrow bandwidth used to eliminate discrete frequency noises, such as the hum from an improperly grounded guitar amp. Notch filters are usually tunable, and can sometimes be used to eliminate specific room or instrument resonances.

octave The musical spacing between a frequency and its double. For example, the distance between "A" (440Hz) and "high A" (880Hz) is an octave. Similarly, the distance between "high A" and "higher A" (1760 Hz) is also one octave. The audible range from 20 Hz to 20,000 Hz encompasses about ten and one-half octaves.

octave band A frequency spectrum which is one octave wide (i.e. all frequencies from 125 Hz to 250 Hz). In recording and audio testing the octave itself is divided into thirds for better accuracy. Each third is named by its center frequency. (63, 125, 250, 1,000, etc.)

omni-directional Equally sensitive to sound from all directions.

oscilloscope An electronic instrument which displays sound waves visually on a scope.

overdubbing The addition of musical or vocal parts to a recording *after* the initial recording session. By the use of overdubbing, all musicians participating in a recording need not be recorded at the same time. This enables one musician to perform more than one part, and, if desired, to do *all* the instruments and vocals in a recording. Generally, an initial "rhythm track" or reference track using drums, bass, piano and guitar is recorded first, and subsequent musical parts are added later. Almost all professional recordings utilize overdubbing to some extent.

passive equalizer An equalizer which does not contain amplification in the equalization circuit. As a result this type of equalizer suffers an inherent "insertion loss," and its output must be amplified, accordingly.

More compact, the *Active* equalizers, incorporating amplification, overcome this problem, and perform the same job with less noise and distortion. As a result, active equalizers are in more common use in console modules, and other studio applications.

patch bay A panel of jacks, corresponding to the inputs and outputs of all devices in the control room. The close grouping of these jacks allows any two devices in the control room to be interconnected by simply inserting a "patch cord" between the appropriate inputs and outputs. If two devices are wired together in the patch bay (the tape outputs to the console inputs, for example), they are said to be "normalled" to one another. The insertion of a patch cord into either input or output jack is said to "break the normal."

phase interference The addition and subtraction of two waves of similar or multiple frequencies, causing peaks and dips in the overall response curve. Phase interference is largely responsible for the "hollow" or "tinny" sound associated with distant miking techniques indoors.

phon The empirical unit of loudness. Since the ear has different sensitivities at various frequencies (Fletcher-Munson), it does not hear equivalent *sound pressure levels* as being equally loud. For example, a 100 Hz tone must be a full *64 dB* (spl) before it is perceived to be as loud as a 1000 Hz tone of only *50 dB* (spl). The phon was created to alleviate this problem, and to measure loudness independently of frequency. Therefore, a 50 phon sound at 100 Hz is perceived to be just as loud as a 50 phon sound at 1000 Hz. For reference, phons are numerically equal to decibels (spl) at 1000 Hz. At other frequencies their value is determined by referring to the Fletcher-Munson Equal Loudness Contours.

pick-up pattern The directional characteristics of a microphone. Standard pick-up patterns are cardioid (heart-shaped), super cardioid, hyper-cardioid, figure eight and omni-directional. Pick-up patterns may differ considerably within one microphone with respect to frequency. Most microphones tend to have more directional pick-up patterns at the high frequencies, because of the smaller wavelengths.

pitch The subjective human perception of frequency. In general, the higher the frequency, the higher the pitch. However, pitch is also a function of the *intensity level* of a sound (owing to the non-linear characteristics of the basilar membrane of the inner ear). For example, if a 100 Hz tone is increased from 60 dB to 100 dB, its pitch goes down by about 10%. This phenomenon also varies with frequency (generally becoming less pronounced at higher frequencies).

plate echo Reverberation produced mechanically by driving thin metal plate with a converted electrical signal. Successive vibrations are picked up by contact microphones along the plate and the resulting simulated reverberation is then mixed with the original signal. The reverberation unit manufactured by EMT utilizes the plate principle.

playback The process of playing back a recorded tape.

punching-in The process of incorporating a "correction" into an otherwise well-recorded track. Depending upon the speed of the recorder's electronics and the engineer's reaction time, single words or notes can be instantaneously overdubbed onto a track without disturbing the continuity of the existing recording.

quadraphonic (Abbreviated: Quad.) The simulation of a 360° listening environment by the proportioning of the audio information between 4 separate channels, each of which is connected to a peripherally placed loudspeaker.

ratio, compression The number of dB that the input signal of a compressor must increase to cause a 1 dB rise in output. (For example, a compressor whose output increases only 2 dB for each 10 dB of output has a 5:1 compression ratio.)

record (verb) To transfer electrical impulses from a sound-producing source to magnetic tape via the process of magnetic induction.

record (noun) A vinyl disc with audio information incorporated into the circumferential grooves on its surface. To be retrieved, the disc must be played with a phono cartridge which converts the impressions in the grooves back to electronic impulses. These impulses can then be amplified and fed to a speaker system.

reflected sound Sound arriving at the listening location after bouncing off one or more of the surrounding surfaces. Because sound waves lose energy according to the distance traveled and number of reflections encountered, reflected sound waves are always of less intensity than similar waves arriving directly from the source. The sum total of all reflected waves determine the room's reverberation time and acoustical character.

reflection The bouncing of a sound wave off of a surface (see *reflected sound*).

resilient Free from rigid contact, such as a spring-mounted floating floor. Resiliency helps prevent the transfer of noise and vibration from one structure to another.

resonance The sympathetic vibration of an object (or air column) at a specific frequency when it is excited into motion by a sound wave of similar frequency in the immediate vicinity.

resonance, sympathetic The vibration of a body or air cavity at its natural resonant frequency, having been set into motion by a nearby vibrating source. Thus a tuning fork that begins vibrating when a nearby fork of similar characteristics is set into motion, is said to be resonating sympathetically. Rattling lamps and buzzing windows are also exhibiting forms of sympathetic resonance. Since these noises are undesirable in recording studios, we must make a special effort to eliminate all potential sympathetic resonators. This includes loose fixtures, hanging lamps and adjacent windows of similar size and thickness in the control room.

resonant frequency dip The degradation of *transmission loss* of a barrier at a specific frequency due to inner resonance. Almost all walls, floors and ceilings exhibit a resonant frequency dip which deteriorates their soundproofing ability. The exact frequency at which this phenomenon occurs is a function of the mass and stiffness of the barrier.

response, frequency The accuracy with which a given device reproduces the audible frequency range. Frequency response is most accurately illustrated by a graph detailing the output of the device at each frequency. In the absence of a graph, frequency response can also be communicated by indicating a range of accuracy, such as "± 3 dB from 20 to 20,000 Hz."

response, transient The response of a device to rapid fluctuations in sound pressure or voltage. Since transients (momentary peaks) occur frequently in music, it is important that a device such as a microphone or amplifier can handle them without distorting.

reverberant field The area in a room, *removed* from the sound source, in which the multitude of decaying reflections has created a reverberant and diffuse condition (see *direct field*).

reverberation The persistence of sound in a room after the actual source has stopped. This is a result of the multiple reflections of sound waves throughout the room arriving at the ear so closely spaced that they are indistinguishable from one another and are heard as a gradual decay of sound.

reverberation (artificial) The simulation of the natural reverberation effect by a mechanical device using springs or plates as the vibrational medium.

reverberation time The time it takes for sounds in a room to decay to one millionth of their original level (60 dB of decay). The reverberation time (T) of a room varies with frequency and is a function of the room volume as well as the total number of absorption units in the room. It is determined approximately by the formula:

$$T = \frac{.049V}{A}$$

where:

T = reverberation time

V = the volume of the room (length x height x width)

A = the total number of units of absorption in the room at a particular frequency (i.e. determined by multiplying the area of each room surface times its absorption co-efficient at a particular frequency).

rhythm track

The initial recorded tracks of a song, serving as the rhythmic foundation for the rest of the piece. The rhythm track usually includes bass, drums, guitar and piano.

root mean square (RMS)

The value of a waveform equal to the value of a continuous signal (i.e. direct current) with the same power output as the initial wave. The RMS value of a *sine* wave is found by squaring all values of the waveform in question and then taking the square root of the average result. This value turns out to always be .707 times the peak value of the wave. The RMS value of a non-sinusoidal wave must take into account all the various frequency components and is more complicated to calculate.

STC (sound transmission class)

A rating system designed to facilitate comparison of the sound transmission characteristics of various architectural materials and constructions. The transmission loss characteristics of a specific material or construction are tested at third octave bands and the results are matched to one of a set of standard STC contours which models the soundproofing ability of that particular construction.

sabin

The measuring unit of absorption; one sabin is equivalent to the absorption offered by a square foot of open window in a room. Sabins are calculated by multiplying the absorption co-efficient of a material times its area.

SelSync The ability of a tape machine to record new material in synchronization with previously recorded tracks, by playing the initial program through selected tracks of the record head. Sel-syncing (also called simul-syncing and sync) enables many parts of one song to be recorded at different times using the one tape and one multi-track recorder.

shifting center An apparent shift of the position of an instrument or voice in the stereo image due to a discrepancy in the phase relationships of the signals from either side. This discrepancy can be caused by electronic processing, such as compression, or by improper room design.

sine wave A waveform whose instantaneous value follows the trigonometric sine function, and whose amplitude variations are as follows: Zero at 0°; Maximum positive value at 90°; Zero at 180°; Maximum negative at 270°; Zero at 360°. A sine wave depicts the variation of amplitude with time of a pure tone (such as that produced by a tuning fork).

sound The phenomenon caused by the vibration of our eardrum. The drum itself is set into motion by pressure waves traveling through the air, originating at the sound source.

sound level meter A pressure-sensitive device which measures loudness. Sound level meters have several scales corresponding to the sensitivities of the human ear at different frequencies.

sound transmission (airborne) The conducting of a sound wave through *air*. The speed of airborne sound transmission varies with temperature and humidity, and is 1130 feet/second in air at 70°F.

sound transmission (structure borne)

The conducting of a sound wave through a *physical structure* (such as a wall, floor, ceiling or door). Because of the increased speeds of sound through common building materials (wood @11,700 feet/second, steel @18,000 feet/second) as well as the physical connection of such materials in the structural framework of a building, structure borne sound transmission is much more difficult to stop than airborne sound transmission, and thus requires special measures to be dealt with effectively.

spring reverb-unit

An artificial reverberation unit utilizing springs as the vibrational medium. Spring units are the least expensive of the artificial reverberation devices, but tend to be the noisiest and most artificial sounding. Thus, they are used where cost is more important than acoustical integrity. This includes electric guitar amps, P.A. systems, and electric organ consoles. Recent technical innovations have allowed some manufacturers to overcome the deficiencies of traditional spring units, and nowadays there are several natural sounding, but slightly more expensive, spring units on the market.

standing wave

A sound wave continuously reinforced by its own reflections. Standing waves produce resonances which alter the natural sound of a room and must be eliminated in both the playing and listening environment. Since the standing waves which form in a room are a direct result of the size and geometry of the space itself, each room has a unique set of standing waves. The presence of these waves can easily be detected by a combination of mathematical calculation and audio analysis and influence the character of all music played or listened to in a room.

tight

A descriptive rather than technical term usually applied to a well defined sound notable for its clarity and distinction. "Tight" usually refers to the absence of excessive reverberation and out of phase reflections.

timbre

The subjective tonal quality of a sound. The timbre of any musical or non-musical sound is determined largely by the *harmonic structure* of the sound wave. Rich sounding musical tones tend to have a great number of inner harmonics which contribute to their lush timbre, while thin sounding musical tones tend to be lacking in the presence of harmonics.

track

A channel or independent division of the tape width, capable of storing information. ½-track stereo recorders divide the tape into two widths, left and right, and are said to have two tracks. 16 and 24 track machines divide the tape into 16 and 24 tracks of information carrying capability.

transient

A momentary peak in program material. Transients are not registered by VU meters, because of their fleeting nature, but are nevertheless capable of producing overload distortion on the tape. (A transient peak can be as much as 14 dB higher than the level indicated by a VU meter.) Percussive instruments, such as drums, bells, and piano, tend to be characterized by many transients in their waveforms. As a result, these instruments must be recorded at slightly lower level than normal to prevent overload distortion.

transmission

The propagation of sound through a medium or barrier (see *sound transmission).*

transmission co-efficient

The portion of sound energy transmitted through a material. A barrier of transmission co-efficient "t" will actually have a transmission loss of $-10 \log t$.

transmission loss (TL)

The number of dB by which a barrier reduces the transmission of sound. If a 100 dB(spl) amplified guitar on one side of a wall was measured to produce only 40 dB(spl) on the other side of the same wall, the TL of the wall would be denoted as 60 dB. Transmission loss varies markedly with frequency, and therefore should be indicated at several frequencies for a given barrier, if an accurate representation of soundproofing ability is to be given.

up front

Refers to a sound notable for its prominence among other sounds.

VU

A visual meter indicating the RMS value of a signal. Since the human perception of loudness corresponds to the RMS value of the signal, VU meters indicate volume (VU stands for Volume Units). Zero VU is considered to be standard operating level.

wavelength

The distance between the beginning and end of a wave or cycle, measured in the vibrational medium. Wavelength is determined by the formula:

$$V = f \times \lambda$$

where:

V = velocity of sound (1130 ft. per second)
f = frequency, Hz
λ = wavelength

Since wavelength and frequency are inversely proportional, low frequencies tend to have much longer wavelengths than high frequencies. For example, a 1000 Hz signal would have a wavelength of approximately 13.5 inches, whereas a 40 Hz tone would have a wavelength over 28 feet in length.

bibliography

Beranek, Leo L. *Acoustics*. McGraw-Hill Electrical and Electronic Engineering Series, McGraw-Hill Book Co, Inc., New York (1954).

——— *Music, Acoustics and Architecture*. John Wiley, Inc., New York (1962).

——— "Sound Transmission through Structures Containing Porous Materials," Ch. 14 in *Noise Reduction,* ed. by Leo L. Beranek, McGraw-Hill Book Co., Inc. (1960).

——— "The Transmission and Radiation of Acoustic Waves by Solid Structures," Ch. 13 in *Noise Reduction* (1960).

——— and R.H. Bolt. "Sound in Small Enclosures," Ch. 10 in *Noise Reduction* (1960).

——— and W.J. Galloway. "Behavior of Sound Waves," Ch. 2 in *Noise Reduction* (1960).

——— and S. Labate. "Properties of Porous Acoustical Materials," Ch. 12 in *Noise Reduction* (1960).

Bolt, R.H., A.D. Frost, S.J. Lukasik, and A.W. Nolle, eds. *Handbook of Acoustical Noise Control,* Vol. I. "Physical Acoustics," Wright Air Development Center, Ohio (Dec. 1952).

Burroughs, Lou. *Microphones: Design and Application.* Sagamore Publishing Company, Inc. New York (1974).

Doelling, N. "Dissipative Mufflers," Ch. 17 in *Noise Reduction* (1960).

Eargle, John, *Sound Recording.* Van Nostrand Reinhold Company, New York (1976).

Franken, P.A. "Reactive Mufflers," Ch. 16 in *Noise Reduction* (1960).

Handbook of Noise and Vibrator Control, ed. by H.R. Warring, Trade and Technical Press Ltd., Morden, Surrey, England (1970).

Helmholtz, H.L.F. *Sensations of Tone* (trans. by A.J. Ellis), Longmans, Green, New York (1930).

"Introduction to Vibration Control Specifications," *VCS Bulletin,* Mason Industries, Inc., New York (Jan. 1, 1973).

King, A.J. "Noise Reduction", Ch. 8, in *Technical Aspects of Sound,* ed. by E.G. Richardson, Elsevier Publishing Co., Amsterdam (1953).

Kosten, C.W. "Behavior of Absorbing Materials," Ch. 4 and "Theory of Sound Transmission," Ch. 5 in *Technical Aspects of Sound* (1953).

Lambert, R.F. "Sound in Large Enclosures," Ch. 11 in *Noise Reduction* (1960).

Mason, Norman J. "Controlling Vibration Problems in Sensitive Structures," ASHRAE Bulletin, Mason Industries, Inc., New York.

McGuinnes, William J. and Benjamin Stein. *Mechanical and Electrical Equipment for Buildings.* John Wiley and Sons, Inc., New York (1971).

Musler, D. and R. Plunkett. "Isolation of Vibrations," Ch. 18 in *Noise Reduction* (1960).

Nisbett, Alec. *The Technique of the Sound Studio.* For Radio, Television and Film. Hastings House, Publishers, New York (1972).

Olson, Harry F. *Music, Physics and Engineering.* Dover Publications, Inc. New York (1967).

Parkin, P.H. and H.R. Humphreys. *Acoustics, Noise and Buildings.* Frederick A. Praeger, New York (1958).

Purcell, J.B.C. "Acoustical Materials for Architectural Uses," Ch. 15 in *Noise Reduction* (1960).

Purkis, H.J. "Room Acoustics and the Design of Auditorea," Ch. 6 in *Technical Aspects of Sound* (1953).

Ramsey, Charles G. and Harold R. Sleeper. *Architectural Graphic Standards.* 6th edition, John Wiley and Sons, Inc., New York (1970).

Rettinger, Michael. *Acoustic Design and Noise Control.* Chemical Publishing Co., Inc., New York (1973).

Richardson, E.G. "Mechanical Musical Instruments," Ch. 18 in *Technical Aspects of Sound* (1953).

Rosenblith, W.A., K.A. Stevens, and the Staff of Bolt, Beranek and Newman. *Handbook of Acoustical Noise Control.* Vol II "Noise and Men," Wright Air Development Center, Ohio (June 1953).

Runstein, Robert E. *Modern Recording Techniques.*

Howard W. Sams & Co. Inc., Indianapolis, Indiana (1974).

Sound Control Construction, United States Gypsum Company, Chicago, Illinois (1972).

Taylor, Rupert. *Noise.* Penguin Books Ltd, Harmondsworth, Middlesex, England (1970).

Tremaine, Howard M. *Audio Cyclopedia.* Howard W. Sams & Co. Inc., Indianapolis, Indiana (1973).

Villchur, Edgar. *Reproduction of Sound.* Acoustic Research, Inc. Cambridge, Massachusetts, Dover Publications, Inc., New York (1965).

index

A

Absorbers
broadband, 47, 131, 164-5
Helmholtz, 46, 131-2
high-frequency, 163
low-frequency (basstraps), 45, 123
membrane, 45, 123 (*see also* Basstraps)
porous, 44-5, 124
temporary, 140
variable, 140-1
Absorption, 43
addition of, 94
Absorption co-efficient, 43-4, 142
of a material, 44
of common building materials (table), 142
Acoustical analysis, 15-16
Acoustical ceiling, 86-7
Acoustical flexibility, 140-2
Acoustic isolation, 132
Acoustic tile, 86-7, 94
Acoustics, room, 28, 151-2
Aesthetics, 116
of the studio, 144-5
Air, sound transmission co-efficient of, 65
Ambience, 59
Ambient noise
common levels of (chart), 30
common misconceptions about, 29-31
defined, 29
in determining soundproofing required, 33-5
and location of studio, 117
maximum acceptable, 29
measurement of, 32-3
Amplitude, 6
Appliances
extraneous noise, 93
"A" scale ("A"-weighted scale), 25
Attack, 14
Audible spectrum, 3

B

Baffles (*see also* Gobos)
hanging, 47
Basstraps, 45, 138, 163
augmentation of, 131
construction of, 125-8
defined, 123
lining of, 124, 126
mechanism of, 123-4
percentage required, 126-8
skewed (multi-dimensional), 129
Bel, 19-20 (*see also* Decibel)
Broadband absorbers, 47, 131, 164-5
Buffer zones, 117
Building permit, 119-20

C

Canopy
for drum booth, 136
Carl F. Eyring Formula, 39
Carpeting, 29, 31
absorption co-efficients of, 142
effects of on soundproofing, 87
on gobos, 136
use in echo elimination, 121
use in soundproofing, 85-6, 87
Ceilings, 84, 102
carpeting of, 85-6
floating, 102-3
single slope, 165
soundproofing of, 84-7
tier drop, 164-6
Coloration, 50
elimination of, 50, 53, 131
as result of standing waves or room resonances, 50
severe, 131
in the studio and control room, 53
Construction material, drywall, 65
Constructive interference, 8
Control room
ceiling, 164-6
coloration in, 53
combined with studio, 149-50
diffusion in, 59-60
dimensional ratios, 153-4
echo in, 55-6
geometry, 154-5
isolation, 152
the optimal, 151
purpose of, 148
size, 153
symmetry, 156-7
visitors area, 166
window, 105-6, 152
Critical distance, 59
Curtains, 29, 31, 156
as variable absorbers, 107, 140
absorption co-efficients of, 142

D

Dead areas, 31, 39-40, 138-40
Decay, 14, 37-8
Decibel, 19
  addition of, 20-1
  converting intensity to, 19
  in human perception of varying frequencies, 23-5
  and pressure (spl), 21
  system, 19
  use of term clarified, 23
Destructive interference, 8
Diffusion, 56-61
  defined, 56
  and direct field, 58-9
  need for, in classical music, 59
  need for, in control room, 59-60
  and reverberant field, 58-9
  and rock music, 59
  in the studio, 59
Dimmers, lighting, 143, 144
Direct field, 58-9
Direct waves, 36
Distortion, harmonic, 12
  production of, diagram, 14
Doors
  in the floating studio, 104
  soundproofing, 67-9
Drum booths, 136-8
  construction of, 136-8
Drums
  isolation of, 136
  leakage, 136
Drywall construction material, 65
Dynamics, internal, 14

E

Echo
  in the control room, 55-6
  defined, 53
  elimination of, 121-2
  flutter type, 54-6, 122
  slapback type, 55
  in small vs. large rooms, 54
  in the studio, 55
Envelope
  of waveform, 14
EQ (see Equalization)
Equalization
  third octave band, 170
  in tuning the room, 169
Equal loudness contours (Fletcher-Munson curves), 23-5
Evaluation
  of the control room, 175
Extraneous noise
  of appliances, 93
Eyring, Carl F., Formula, 39

F

Finish materials, 142
  absorption co-efficients of, 142
Fletcher-Munson curves (equal loudness contours), 23-5
Floating studio, 95-107
  defined, 97
  electrical wiring in, 110-13
  ventilation, 107-10
Floor constructions
  soundproofing abilities, 88-92
Floor plans
  combined studio and control room, 150
  control room, 154-7
  studio, 118-19
Floors
  floating, 100-2
  soundproofing of, 87-92
Flush mounting of monitors, 160-2
  advantages of, 162
Flutter echo, 54-6, 122
  control of, 122
Frequency
  defined, 3
Frequency analysis
  by third-octave bands, 15-16
Frequency response
  in balance, 153
Frequency-spectrum graph, 15-16
Fundamental tone, 8

G

Geometry
  of control room, 154-5
Gobos
  defined, 132
  design and construction of, 132-3, 134-5
  size of, 134
  surfacing of, 134
  why used, 132

H

Harmonic distortion, 12
  production of, diagram, 14
Harmonics, 8-9
  adding acoustical color, 10
  defined, 8
  shaping waveform, 11-12
Hearing, threshold of, 19
Helmholtz resonators, 46, 131-2
Hum, 60-cycle, 168

I

Intensity, 18, 20
  effect of doubling, in dB, 20
  measurement of, 18
  relationship to pressure, 21
Interface, 167-8
  and use of troughs, 168-9
Interference
  constructive and destructive, 8
  phase, 161
  radio frequency, 169
Internal dynamics, 14
Isolation
  and control room, 152
  of the drums, 136
  of instruments, 41
  of studio, 117

L

Larger budgets, soundproofing within, 64, 95-113
Layout
  of studio, 118-19
Leakage, 41
Lighting, 143-5
  natural, 107, 145
Live areas, 36, 39-40, 138-40
Logarithmic scale, 15
Loudness
  apparent or perceived, 17-18, 30
  correspondence to amplitude, 6
  RMS (root mean square), defined, 17
Loudspeakers (*see* Monitors)

M

Machine soffits, 162
Measurements
  of ambient noise, 32-5
  of monitor/room response, 170
  of reverberation, 39
  of sound pressure, 21-3
Membrane absorbers, 45
Meter, sound-level, 21, 32
  and "A"-weighted scale, 25
  use of, in determining amount of equalization necessary, 170
Monitor path, 160
Monitors
  aiming and positioning of, 156-9
  and assessment of sound, 59-60
  directional performance of, 60
  flush mounting of, 160-2
  and monitor path, 160
  mounting of, in combined studio/control room, 149
  response of, 170
  spacing of, 158
  tuned to the room, 169-71
Musical analysis, 15
Musical instruments
  differences in timbre, 10
Musical sounds
  distortion, 12
  harmonics, 8-9

N

Noise, ambient
  common levels of (chart), 30
  common misconceptions about, 29-31
  defined, 29
  and determining soundproofing required, 33-5
  and location of studio, 117
  maximum acceptable in a studio, 29
  measurement of, 32-3

O

Octave bands, 15-16
Octaves, 9
Openings, room, 65
  sealing of, 65-6

P

Panel absorbers, 45 (*see also* Absorbers, membrane)
Parallelism, 120
  and avoiding it, 120-1
  and flutter echo, 54, 122
Permit, building, 119-20
Phase, 7 (*see also* Reflections, out of phase)
Phase cancellation and reinforcement
  by tier ceiling, 166

Phase interference, 161
Phase relationship, 8
Porous absorbers, 44-5
Proportions
  of basstrap to room surface area, 126-8
  of control room dimensions, 153-4
  for skewing walls, 120

R

Radio frequency interference, 169
Ratios, room (*see also* Proportions)
  acceptable, 153
Rear radiation, 162
Reflections, rear room, 163-4
  out of phase, 163
Resonance, sympathetic, 16, 93 (*see also* Coloration)
Reverberant field, 58-9
Reverberation, 36 (*see also* Coloration)
  formation of, 36-7
  and live vs. dead rooms, 39-40
  measurement of, 39
Reverberation time
  computation of, 38-9
  how generally used, 43
  optimal, 41-3
RMS (root mean square), 17
Room acoustics, 28, 151-2
Room openings, 65
  sealing of, 65-6
Rooms, live vs. dead, 39-40
Room tuning, 169-73
Root mean square (RMS), 17

S

Shape, wave, 6, 11-13
Size
  of control room, 153-4
Skew
  of ceilings, 165
  of walls, 120-1
  of window panes, 105
Slapback, 55
Small budgets, soundproofing within, 64, 65-94
Smearing, 175
Soffits, machine, 162
Sound
  defined, 3
Sound-level meter, 21, 32
  and "A"-weighted scale, 25
  use of, in determining amount of equalization necessary, 170
Sound-level readings
  in direct field, 58
  in reverberant field, 58-9
Sound locks, 104
Sound pressure, 21
  measurement of, 21-3
Soundproofing for larger budgets, 95-113
  ceilings, floating, 102-3
  doors, 104
  floors, floating
    concrete, 101-2
    wood, 100-1
  ventilation, 107-10
  walls, floating, 97-9
  windows, 105-7
  wiring, 110-13
Soundproofing for small budgets, 65-94
  appliances, 93-4
  ceilings, 84-7
  doors, 67-9
  floors, 87-92
  loose objects, 93
  room openings, 65-6
  walls, 72-83
  windows, 69-72
Sound seal, 152
Sound transmission co-efficient
  of air, 65
Speakers (*see* Monitors)
Spectrum, audible, 3
Standing waves
  axial, 50
  defined, 50
  detection of, 132
  elimination of, 50, 53, 123
  formation of, 50-1
  oblique, 52
  and skewed walls, 120-1
  in the studio and control room, 53
  tangential, 52
Sustain, 14
Symmetry
  of control room, 156-7

T

Tape machines
  placement of, 149, 154-5
  soffits, 162
Temporary absorbers, 140

Third-octave bands, 15-16
Third-octave equalization, 170-1
  problems with, 171-3
Threshold of hearing, 19
Tier-drop ceiling, 164-6
Timbre, 10-11
TL (transmission loss), 33, 65, 67
Transmission loss (TL), 33, 65, 67
Tuning, room, 169-73

V

Variable absorbers, 140-1
Velocity of sound, 4-6
Ventilation, 107-10
Visitors' area, 166

W

Wall constructions
  soundproofing abilities, 73-83
Walls
  floating, 97-9
  soundproofing of, 69-72
Waveform
  complex, 6
  energy content of, 17
  envelope, 14
  sections of, 14
  simple, 6
Wavelength, 5-6
Waves, direct, 36
Waves, reflected, 36-7
Waves, sound
  defined, 3
Waves, standing
  axial, 50
  defined, 50
  detection of, 132
  elimination of, 50, 53, 123
  formation of, 50-1
  oblique, 52
  and skewed walls, 120-1
  in the studio and control room, 53
  tangential, 52
Weighted scales
  the "A"-weighted scale, 25
Window, control room, 105-6, 152
Windows
  soundproofing of, 69-72
  in the studio, 107
Wiring, 110-13, 167-9

about the author

Jeff Cooper is one of the leading authorities on recording studio design and construction in the world. Born in Toronto, Canada, he studied architectural acoustics at the *Massachusetts Institute of Technology* in Cambridge, Mass., where he was awarded a Master's degree for his work on recording studio acoustics. His mentor and thesis advisor was Robert Newman of the internationally known acoustical consulting firm of *Bolt, Beranek and Newman.*

Jeff Cooper received three additional degrees from MIT, including a second Master's degree in project management from the *Sloan School of Management*, by the age of 22.

In 1976, after working as a designer and assistant to the president at *Westlake Audio*, Mr. Cooper started an acoustical consulting and architectural firm, *Jeff Cooper Architects, A.I.A.*, based in Los Angeles, to specialize in the design of recording studios and projects for the performing arts.

His clients cover a wide span, ranging from members of well known rock bands *(America, Chicago)* seeking private studios, to international record companies *(Polygram, Capitol, EMI Records)* seeking state of the art studio complexes.

In 1977, Jeff Cooper was chosen to design a revolutionary computerized film mixing studio for the renowned producer and director, *Francis Ford Coppola*. With the success of this installation, much film related work followed, and Mr. Cooper has recently completed film dubbing centers and theaters for *George Lucas* (Star Wars) and *Steven Spielberg, MCA Universal, Ryder Sound, Warner Bros., and others.*

Mr. Cooper's firm has developed many innovative techniques in the area of architectural acoustics, including *the Kinetic Diffuser Slat System™, the Broadband Acoustical Diffuser™* and *the Infinite Baffle Speaker Wall™*, recently adopted in *Lucasfilm's THX™* theater sound system.

As both a licensed architect and an acoustical engineer, Mr. Cooper has now designed over 150 projects throughout the world. His work is represented in Europe, Australia, South America and Asia, as well as the United States and Canada.

As a speaker, Jeff Cooper has addressed the *Recording Institute of America*, the *Audio Engineering Society*, and the *Acoustical Society of America*.